PU[illegible] BOOKS

MATHS SUTRAS FROM AROUND THE WORLD

Gaurav Tekriwal is the founder president of the Vedic Maths Forum India. An educator, Gaurav has been imparting Vedic mathematics skills over the past eighteen years across the globe. He inspires and informs people, helping them to realize their true potential by introducing them to the world's fastest mental-maths system—Vedic mathematics.

Gaurav is the author of *Maths Sutra – The Art of Speed Calculation*, *Speed Math* and *Speed Calculation* (In Hindi).

Through television programmes on the Tata Sky and Reliance Big TV platforms, DVDs, books and online courses he has taken the Vedic maths system to over four million students in India, South Africa, the United States, Australia, UAE, Ghana and Colombia. Gaurav is a five time TED/TEDx Speaker. Named one of India's Top Young Visionaries by Indiafrica and Ministry of External Affairs of India, he was awarded by the Honorable Governor of West Bengal for his contribution to the field of Vedic mathematics in 2016.

For more information, please visit www.mathssutra.com or www.vedicmathsindia.org

MATHS SUTRAS from around the World

Speed Calculations on Your Fingertips

GAURAV TEKRIWAL

Illustrations by Jit Chowdhury

PUFFIN BOOKS

An imprint of Penguin Random House

PUFFIN BOOKS

USA | Canada | UK | Ireland | Australia
New Zealand | India | South Africa | China

Puffin Books is part of the Penguin Random House group of companies
whose addresses can be found at global.penguinrandomhouse.com

Published by Penguin Random House India Pvt. Ltd
7th Floor, Infinity Tower C, DLF Cyber City,
Gurgaon 122 002, Haryana, India

First published in Puffin Books by Penguin Random House India 2017

10 9 8 7 6 5 4 3 2 1

ISBN 9780143333852

Typeset in Adobe Garamond Pro by Manipal Digital Systems, Manipal
Printed at Thomson Press India Ltd, New Delhi

www.penguin.co.in

This is for you Miraaya –
My darling three-year-old!

Contents

In this chapter, we will see how to check our calculations in addition, subtraction and multiplication with one single method called the 'digit sum check'. As we go along, we see the concept of casting out nine and the 9-point circle. You will also get hands-on activity to find out the patterns which exist in the 'Vedic square'. Overall, be amazed at the connection within maths and the existing patterns!

There are plenty of methods to make fractions easier. In this chapter, we again explore the 'Singapore maths' way of doing word problems for fractions, based on the visual bar model method. The Vedic maths way comes in handy when it comes to addition and subtraction of fractions, where we use the maths sutra, vertically and crosswise.

If you would like to square numbers mentally, this chapter is just for you. We first learn the squaring of numbers ending in 5; then we learn the squaring of numbers close to a base and finally we learn a general method of squaring two-digit figures. This chapter comes to life with various activities and as a surprise, you also get to learn about the 'magic of magic squares', from the life and times of the Indian mathematician Varahamihira, who lived around 550 AD.

Preface

This book is for young and enthusiastic learners and explorers of mental arithmetic. It covers methods of maths, which can also be called maths sutras, from around the world–from India to Egypt, from Singapore to Japan. You will also get numeorus activities and worksheets that will give you a better grip of the methods shown. Overall, the objective of this book is to make you think mentally and make maths entertaining and enjoyable.

You must be wondering about the need for mental maths methods in this age of calculators and technology? Well, our brain behaves like a muscle. Just like the body, it needs regular exercise too. If you practise regular mental maths exercises and puzzles, you will not only have a sharper mind for math but also sharper logical reasoning skills.

I must also add a that currently, numeracy skills are going down worldwide and not many people can even do basic maths. There is also a phobia around maths. Students loathe it. And in such a scenario, to excel in maths becomes very challenging. That's where the methods discussed in this book will add value by side-stepping the traditional system and making maths doable and ultra-cool to boot.

Over the last fifteen years, in my passion and quest for simple and quick maths methods for students, I have looked deeply into the concepts of Indian Vedic mathematics, rediscovered by the Indian saint, Tirthaji. I have found that it works beautifully with students across the globe. My thirst for more such methods in other cultures led me to conceptualize this book for young learners. As I explored, I was amazed to discover the learning methods of people in other countries. For example, in Singapore, they have devised a system to do word problems in a visual way called the bar modelling technique. These systems have been included in the curricula in many countries, including the United States of America. When it comes to maths skills, Singapore also ranks number one in the world in a global study of learners.

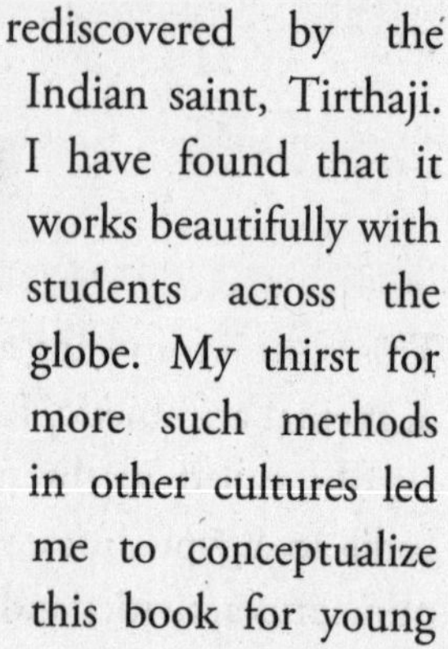

Japan has an amazing grid puzzle culture. Sudoku has been a worldwide sensation and appears in hundreds of newspapers around the planet. There are other puzzles too, like *Kakuro* and *KenKen*, which fascinate me. Students who use this book will benefit greatly from these puzzles which, while being fun-filled, help improve one's mathematical reasoning and logical thinking skills. Watch out for KenKen championships in your city as they get popular and get on top of their puzzle leader board.

Egypt, the name itself takes us back several millennia to the land of the Pharaohs and Pyramids. I was blown away by a YouTube video by Michael S. Schneider, mathematician-author-educator on Egyptian maths. It was so simple and striking that I used some of the content in this book, so that it reaches the right audience—you! No more struggling with traditional methods. I sometimes wonder whether one day my child would be able to learn the best from all civilizations! Maybe she can someday sit in a classroom and learn, say multiplication from India, division from Egypt, and fractions from Singapore. Wouldn't that be cool? Maybe we can even have a global maths curriculum incorporating the best of all worlds for everybody.

And yes, no more rote learning, no more maths phobia, and certainly, no more tears. Maths can be fun too and this book shows you exactly that! In short, in this book you will see invisible connections between the methods and at the end, take home an endless enchantment with numbers. So without any further ado, let's get the maths magic started!

1

All About 'Additions'

'The only way to learn mathematics is to do mathematics.'

—Paul Halmos

You must have done addition since you were four or five years old. It is one of the most understood topics that most young people like you excel in. I remember when I was your age, I used to love doing addition problems, I found them rather easy. My teacher at school used to give me A+ with five stars and a sticker whenever I answered a set of problems on addition correctly.

Apparently one day when I was given a set of problems on addition to do mentally, I struggled. I was asked to add two digits to another two digits in a problem—it was somehow difficult. And I found bigger digits, say a three-digit number plus another three-digit number even more cumbersome and challenging to do mentally. It was all right to solve it on paper but my mind drew a blank when I was asked to do the same problem mentally. It was scary!

In our schools, mental mathematics is given due importance but there aren't any fixed rules given about how to approach a problem orally—be it addition or division, for that matter. The oral system is taught nowhere because we give importance to step-by-step calculation. I don't refute that but I feel we must also be taught the oral way of maths, which comes in handy while doing the four operations—addition, subtraction, multiplication and division—mentally as you will see in this book.

Doing the four operations and more in less time gives one a sense of accomplishment and a solid, robust mental framework in calculations. It adds to your creativity in the way you do maths. Any problem can be solved mentally in more than one way, as I will show you now.

So, let's get started with addition and learn how to do it mentally.

Mental Additions

Sometimes, we find it difficult to add numbers which end in 6, 7, 8 and 9. For example, if we have 16 + 9, that's a difficult problem to do mentally.

But we can make it easy. We can use a method called 'by addition and by subtraction'.

Let us try 16 + 9.
Since adding 9 directly is difficult for most of us, we add 10 which is easy to do—mentally. So, since 9 is 1 less than 10, we can add 10 and then subtract 1 from our answer.
Our sum looks like this:

$$16 + 9$$

We do: 16 + 10 = 26
26 – 1 = 25 is our answer.
Here, I would like to draw your attention to the method which is called 'by addition and by subtraction'. So, we add first and then subtract. Hope this bit is clear to you.

Let's take another example. Say, we have:

68 + 9

We do 68 + 10 = 78.
78 – 1 = 77 our answer.

Now, let's try another example on our own.

59 + 8

We do: 59 + 10 = 69
Since 8 is 2 less than 10, we do:
69 – 2 = 67. The answer is 67.

Now can you tell me what happens if one of the numbers being added was closer to 20?
Say, we have 116 + 18
So, to add 18, we add 20 and then subtract the 2. That was quite simple!
So, we do:
116 + 20 = 136
136 – 2 = 134. This is our answer.

Now let's try addition with a larger number
139 + 69

Here, to add 69, we add 70 and then subtract 1 from the total.
139 + 70 = 209
209 – 1 = 208 our answer.

Let's try another problem.
166 + 88
To add 88, we added 90 and then subtract 2.
166 + 90 = 256
256 – 2=254 was the answer.

ACTIVITY 1

Find the sum.

1. 88 + 7 = ____

2. 68 + 6 = ____

3. 59 + 9 = ____

4. 42 + 9 = ____

5. 53 + 9 = ____

6. 27 + 5 = ____

7. 95 + 7 = ____

8. 56 + 8 = ____

9. 49 + 8 = ____

10. 19 + 6 = ____

Find the sum.

11. 129 + 98 = ____

12. 954 + 86 = ____

13. 167 + 89 = ____

14. 224 + 67 = ____

15. 375 + 88 = ____

16. 942 + 89 = ____

17. 886 + 95 = ____

18. 428 + 79 = ____

19. 774 + 78 = ____

20. 215 + 99 = ____

Left to Right Mental Additions

Let us now see another method of quicker addition. This method is called the 'left to right mental addition' method. So far, traditionally in maths, we have been doing additions and other operations from right to left. Now, in this method, we will get our answers efficiently from left to right, hence disrupting and sidestepping the traditional calculating system.

Say for example, we have 78 + 45.

$$\begin{array}{r} 78 \\ +\ 45 \\ \hline \end{array}$$

Step 1

We first add the figures in the left column. So 7 + 4 = 11. We keep that figure in our head.

$$\begin{array}{r} 78 \\ +\ 45 \\ \hline 11, \end{array}$$

Step 2

We then add the figures in the right-hand column 8 + 5 = 13. We keep that in our head too. The sum looks like this.

$$\begin{array}{r} 78 \\ +\ 45 \\ \hline 11,13 \end{array}$$

Step 3

In our final step, we add (or combine) the middle digits.

$$\begin{array}{r} 78 \\ +\ 45 \\ \hline 11,13 \\ 123 \end{array}$$

So, the answer is 123. I hope this is clear.

Let's take another example to understand it better.
Say we have 87 + 38.

$$\begin{array}{r} 87 \\ +\ 38 \\ \hline \end{array}$$

Step 1

We add the figures in the left column 8 + 3 = 11. We get 11, we keep that figure in our head.

$$\begin{array}{r} 87 \\ +\ 38 \\ \hline 11 \end{array}$$

Step 2

We then add the figures in the right-hand column 7 + 8 = 15. We get 15. We keep that in our head too. The sum looks like this now:

$$\begin{array}{r} 87 \\ +\ 38 \\ \hline 11,15 \end{array}$$

Step 3

In our final step, we add (or combine) the middle digits.

$$\begin{array}{r} 87 \\ +\ 38 \\ \hline 11,15 \\ 125 \end{array}$$

So, the answer is 125.

I hope you have understood how to add two-digit numbers. But what if we have to add two three-digit numbers? There is a solution for that as well.

Say we have to add 582 + 759

$$\begin{array}{r} 582 \\ +\ 759 \\ \hline \end{array}$$

Step 1

Start by adding the columns from left to right. The first column is 5 + 7 = 12. The middle column is 8 + 5 = 13.

$$\begin{array}{r} 582 \\ +\ 759 \\ \hline 12,13 \end{array}$$

Step 2

We add (combine) the middle digits of the first two columns. So, we have 133 in our head.

Then we add the last column on the right. So, we have 2 + 9 = 11. Now the sum looks like this:

$$\begin{array}{r} 582 \\ +\ 759 \\ \hline 133,11 \end{array}$$

Step 3

So, in our mind, we have 133, 11. We then combine or add the digits on either side of the comma. In this case, we add 3 and 1 and get 4. The final answer is 1341.

Let's take another example. Say, we have 698 + 576

$$\begin{array}{r} 698 \\ +\ 576 \\ \hline \end{array}$$

Step 1

We add the first two columns from left to right. We get 11, 16.

$$\begin{array}{r} 698 \\ +\ 576 \\ \hline 11,16 \end{array}$$

Step 2

We add (combine) the first two columns. So, we have 126 in our head. We then total up the final column and get 8 + 6 = 14. Finally our example looks like this:

$$\begin{array}{r} 698 \\ +\ 576 \\ \hline 126,14 \end{array}$$

Step 3

We combine the digits on either side of the comma. So, in this case we add 6 + 1 = 7.

$$\begin{array}{r} 698 \\ +\ 576 \\ \hline 126,14 \\ 1274 \end{array}$$

Our final answer is 1274.

ACTIVITY 2

Find the sum.

1. $\begin{array}{r} 63 \\ +\ 89 \\ \hline \\ \hline \end{array}$ **2.** $\begin{array}{r} 78 \\ +\ 58 \\ \hline \\ \hline \end{array}$ **3.** $\begin{array}{r} 95 \\ +\ 59 \\ \hline \\ \hline \end{array}$ **4.** $\begin{array}{r} 56 \\ +\ 49 \\ \hline \\ \hline \end{array}$

5. $\begin{array}{r} 78 \\ +\ 35 \\ \hline \\ \hline \end{array}$ **6.** $\begin{array}{r} 82 \\ +\ 29 \\ \hline \\ \hline \end{array}$ **7.** $\begin{array}{r} 68 \\ +\ 45 \\ \hline \\ \hline \end{array}$ **8.** $\begin{array}{r} 86 \\ +\ 44 \\ \hline \\ \hline \end{array}$

9. $\begin{array}{r} 83 \\ +\ 39 \\ \hline \\ \hline \end{array}$ **10.** $\begin{array}{r} 75 \\ +\ 76 \\ \hline \\ \hline \end{array}$

Find the sum.

11. $\begin{array}{r} 575 \\ +\ 697 \\ \hline \\ \hline \end{array}$ **12.** $\begin{array}{r} 489 \\ +\ 773 \\ \hline \\ \hline \end{array}$ **13.** $\begin{array}{r} 238 \\ +\ 589 \\ \hline \\ \hline \end{array}$ **14.** $\begin{array}{r} 795 \\ +\ 729 \\ \hline \\ \hline \end{array}$

15. $\begin{array}{r} 686 \\ +\ 975 \\ \hline \\ \hline \end{array}$ **16.** $\begin{array}{r} 199 \\ +\ 832 \\ \hline \\ \hline \end{array}$ **17.** $\begin{array}{r} 939 \\ +\ 985 \\ \hline \\ \hline \end{array}$ **18.** $\begin{array}{r} 868 \\ +\ 954 \\ \hline \\ \hline \end{array}$

19. $484 + 357$ = ______

20. $718 + 775$ = ______

Find the sum.

21. $8,763 + 9,653$ = ______

22. $5,696 + 8,197$ = ______

23. $5,695 + 9,017$ = ______

24. $6,965 + 6,398$ = ______

25. $9,390 + 9,966$ = ______

26. $2,519 + 7,579$ = ______

27. $9,449 + 9,238$ = ______

28. $3,754 + 6,745$ = ______

29. $5,852 + 9,956$ = ______

30. $3,010 + 6,849$ = ______

> 'Mathematics is a language plus reasoning; it is like a language plus logic.
> Mathematics is a tool for reasoning.'
>
> —Richard P. Feynman, 1965 Nobel laureate

So that was mental addition made easy for you. These methods and systems work elegantly if learnt and practised well. I hope you now have a solid grounding in addition. But to take things little further and to make addition really cool, I would like you to dabble with puzzles on addition from the land of the rising sun—Japan. These puzzles will make your logical thinking and mathematical reasoning skills strong.

So, did you know that Japan has given the world giants like the Sony Corporation, sumo wrestling, Godzilla and tsunamis? And did you know that if you were to rank countries with their maths scores, Japan ranks among the world's best! (Source: *The Trends in International Mathematics and Science Study* [TIMSS] 2015 Global Study).

So, if you dig deeper into this and scout for reasons behind Japan's success in maths, you can be sure that it can be attributed to the amazing logical puzzles created by some amazing Japanese mathematicians. Take Kakuro, for instance.

The name Kakuro is Japanese for *kasan kurosu*, meaning 'addition cross'. Kakuro puzzles have been featured daily in the *Guardian*, the *Daily Mail*, the *New York Times* and many other publications. The Japanese love their puzzles and I can say that Kakuro is the second most popular puzzle in the country. And it is catching on in other countries too because of the way it helps improve mathematical reasoning and logical thinking skills. So, let me share with you a 3 × 3 Kakuro puzzle and the rules on how to solve it. We will start small and slowly expand to bigger Kakuro puzzles.

Rules: Just like letters in crossword puzzles, Kakuro puzzles are to be filled with figures from 1 to 9, each figure to be used only once in a particular entry. You should ensure that no digit is duplicated in any entry. You may have repetitions of a digit in a different entry—that's absolutely fine and you will encounter many repetitions in 4 × 4 or bigger Kakuro puzzles. You need to ensure that these figures add up to the numbers mentioned on the puzzle. You must also remember that each Kakuro puzzle will have a singular solution. All the best and have an awesome number-crunching experience!

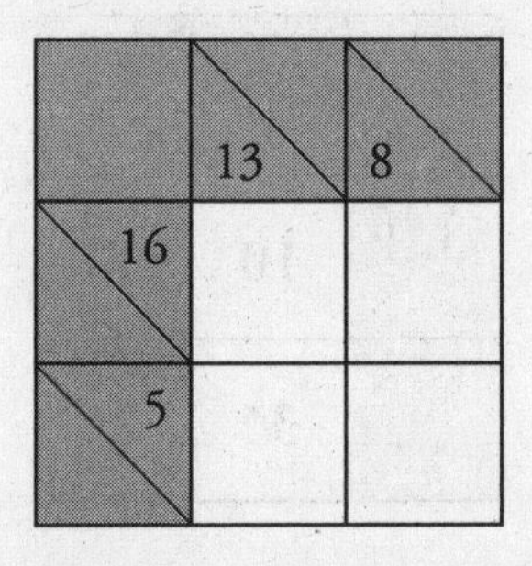

Kakuro puzzles will contain many clue squares; these are squares which help you to solve the puzzle. A clue square can have an 'across' clue or a 'down' clue, or both.

Step 1

Let's see step number one. So, start from the lowest number 5. So, we see that we have 5 across.

So, our options are either 1 + 4 or 2 + 3.

	13	8
16	**11**	**5**
5	**2***	**3***

Now for example, if we use 2 and 3 across, it would be invalid. Because then it would be vertically 2 + 11 and the puzzle will look like this.

Even though this fits, we can't use 11 as we have to use digits between 1 to 9. So, we change it and put 3 and 2.

	13	8
16	**10**	**6**
5	**3***	**2***

Again, we can't use 3 as we then need 10+3 to make 13. We can't use 10 as the digits need to be between 1 to 9. So, this solution below is invalid as well.

Step 2

So what else adds to 5? How about 4+1?

So, we take the other option 4 and 1 to add up to 5. Then for 13 down, we would need 9 which is correct. And then for 8 down, we have 7 and 1. Since 1 is already there we can now say that we have our unique solution.

Our completed puzzle looks like this:

	13	8
16	**9**	**7**
5	**4**	**1**

Easy enough? Hope you liked this. You will love it more when you solve a Kakuro puzzle on your own. So, let's go ahead and get you initiated. In Activity 3 below, I have given six Kakuro Puzzles and also some hints in each of them. Try them now!

ACTIVITY 3

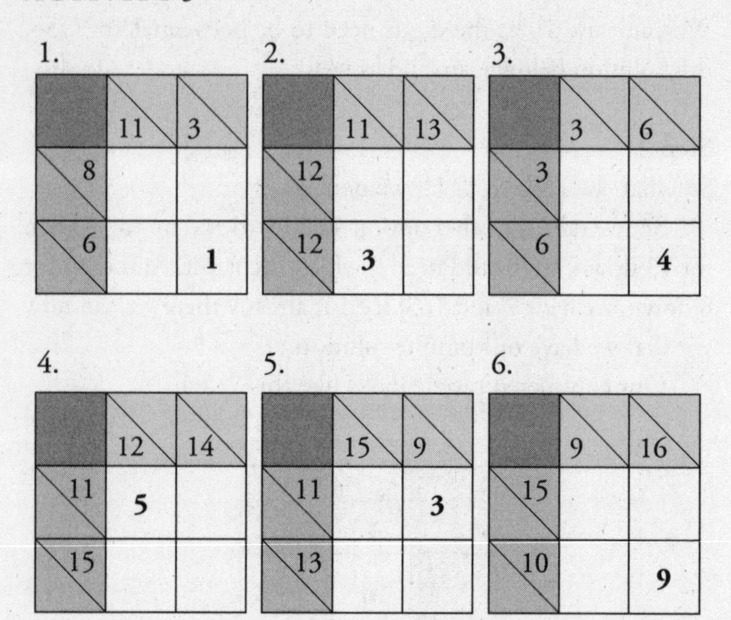

Now Kakuro puzzles can be of 3 × 3 or even 4 × 4 or even more than 30 × 30 as well. I am giving you a few 4 × 4 Kakuro puzzles below as Activity 4, so that you get to understand the basic concept behind it and then are able to do Kakuro puzzles of any size. And to make it simpler for you, I have given hints too. So, let's get started!

ACTIVITY 4

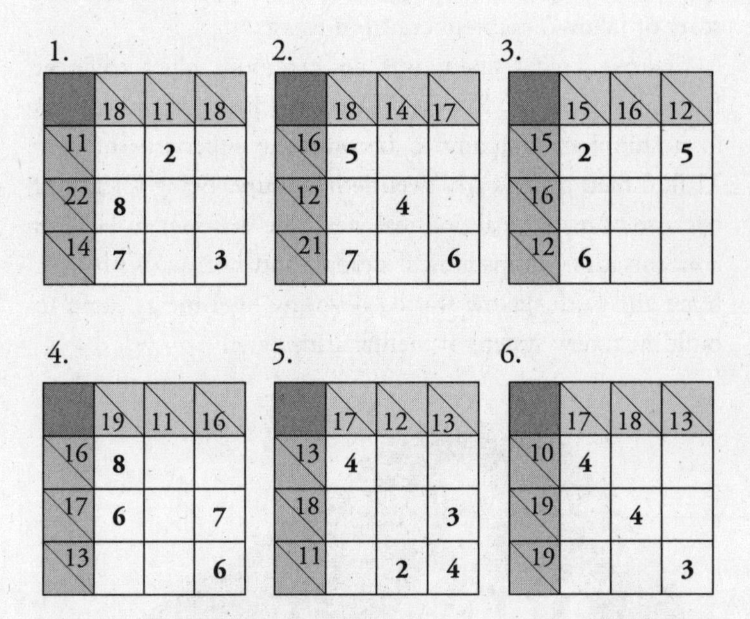

I hope you saw mental additions in a new light and found Kakuro riveting! Just Google Kakuro and you will find plenty of websites offering these puzzles for practice for free. You could also visit www.kakuroconquest.com.

Before we go forward, I must share with you the inspiring story of Jakow Trachtenberg from Russia.

Jakow Trachtenberg was an ingenious chief engineer from St Petersburg, Russia. He quickly climbed up the ranks in a shipping company to become the supervisor of over 11,000 men in the early twentieth century. Being a Jew and because of his radical political views, he was sent to Hitler's concentration camps amidst despair and horror. With little hope and faith, Jakow started devoting his time at camp to building a new system of mental arithmetic.

He had no paper, pencil or pen. Simply sitting in the camp, he found new ways to look at numbers—mentally. Later, Jakow escaped camp after spending years in it and founded the Mathematical Institute in Zurich, Switzerland. Jakow believed and concluded that people are born with 'phenomenal calculation possibilities'.

And this is why I wanted to share the story with you to urge you to believe in yourself—no matter what the circumstances!

> 'We must accept finite disappointment, but never lose infinite hope.'
>
> —Martin Luther King Jr

Now, let me share with you Jakow Trachtenberg's method of columnar addition.

High Speed Columnar Addition

Let's take a sum. Say we have: 1256 + 8563 + 4658 + 2387 By the conventional method, we would have added from the rightmost column, going 6 plus 3 plus 8 and so on. But in this system, we can begin by working on any column. So, let's start from the leftmost column.

```
  1256
  8563
  4658
+ 2387
  ----
```

Step 1

Here the rule is that we *'never count more than 11'*. We do the addition and whenever the running total becomes greater than 11, we reduce it by 11 and go ahead with the reduced figure called the *running total.* And as we do so, we make a small tick or check mark beside the number that made our total higher than 11.

So, we have, from the leftmost column 1 + 8 = 9 + 4 = 13.

Now this is more than 11. So, we subtract 11 from 13. Make a tick and start adding again with 2. We go on and add 2 so we get a running total of 4.

We write this under the column next to running total.

And next to ticks we write 1. Our example now looks like this:

	1256
	8563
	4658
	+ 2387
Running Total	4
Ticks	1

We now go on to the next column from left and complete it too.

So, we go 2 + 5 = 7 + 6 = 13.

Therefore 13 is more than 11, so we put a tick and our running total becomes 2.

Then we go on 2 + 3 = 5.

So, our running total becomes 5 and number of ticks is 1.

Our sum at the end of all the four columns looks like this:

	1256
	8563
	4658
	+ 2387
Running Total	4 5 2 2
Ticks	1 1 2 2

Step 2

Now we arrive at the final result by adding together the running total and ticks in this way. We add in the shape of the alphabet 'L'.

		1256	
		8563	**Add in shape of L**
		4658	
	+	2387	
Running Total		4 5 2 2	
Ticks	0	1 1 2 2	0

So, we have 2 plus 2 plus 0 = 4 in the unit's place.
Then we have, 2 + 2 + 2 = 6 in the ten's place.
Then we have, 5 + 1 + 2 = 8 in the hundreds place.
Then we have, 4 + 1 + 1 = 6 in the thousands place.
Finally, we have, 0 + 0 + 1 = 1
So, our answer becomes 16864
Easy? This is an alternative method to do quick columnar additions. Let's go ahead and take another sum as our example.
Say we have 89621 + 46892 + 14780 + 45893 + 29003

$$
\begin{array}{r}
89621 \\
46892 \\
14780 \\
45893 \\
+\ 29003 \\
\hline
\end{array}
$$

Step 1

We start doing the sum from the leftmost columns. We never add more than 11 as a thumb rule. We get our running total and ticks as follows:

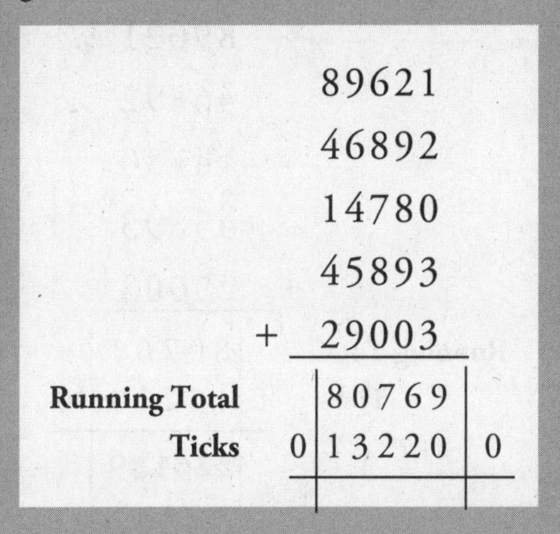

Step 2

We now add in the shape of the alphabet 'L'.

For the units place we get 9 + 0 + 0 = 9

For the tens place, we get 6 + 2 + 0 = 8

For the hundreds place, we have 7 + 2 + 2 = 11. We put 1 down and carry-over 1.

For the thousands place, we have 0 + 3 + 2 + 1 (carry-over) = 6

For the ten thousands place, we have 8 + 1 + 3 = 12. We put 2 down and carry-over 1.

For the lakh's place, we have 0 + 0 + 1 + 1 (carry-over) = 2.

		89621	
		46892	
		14780	
		45893	
	+	29003	
Running Total		80769	
Ticks	0	13220	0
		226189	

So, our answer becomes 226189.

2

The 'Difference' Matters

In the last chapter we learnt how to add mentally with the method 'by addition and by subtraction'. In this chapter on subtraction, we will be doing just the opposite. So, we will apply the rule 'by subtraction and by addition' in the case for mental subtractions.

Mental Subtractions

Sometimes it becomes difficult to subtract numbers like 6, 7, 8, 9 from numbers ending in 1, 2, 3, 4, 5. This new method provides us with a different view and simpler approach, just like we saw in addition.

Say, we have 42 – 9

We subtract 10 first (as it is easy to take away 10) and then add back 1 (since 9 is 1 less than 10).

42 – 10 = 32

So, 32 + 1 = 33. This is the answer.

Let us take another example:

71 – 8

So, we subtract 10 first (as it is easy to take away 10) and then add back 2 (since 8 is 2 less than 10).

71 – 10 = 61

So, 61 + 2 = 63 is the answer.

Now let's try one with a slight variation!

53 – 17

So, we subtract 20 first (as it is easy to take away 20) and then add back 3 (since 17 is 3 less than 20).

53 – 20 = 33

So, 33 + 3 = 36 is the answer.

Let's try another variation.

172 – 38

Here, we will first subtract 40 and then add 2.

172 – 40 = 132

So, 132 + 2 = 134 is the answer.

Let's try another one!

541 – 67

We subtract 70 and then add 3.

541 – 70 = 471

471 + 3 = 474 is the answer.

ACTIVITY 1

Find the difference.

1. 17 − 8 = ____
2. 57 − 9 = ____
3. 63 − 6 = ____
4. 34 − 6 = ____
5. 24 − 8 = ____
6. 33 − 7 = ____
7. 63 − 6 = ____
8. 12 − 8 = ____
9. 94 − 6 = ____
10. 37 − 9 = ____

Find the difference.

11. 122 − 59 = ____
12. 411 − 78 = ____
13. 251 − 69 = ____
14. 193 − 88 = ____
15. 865 − 47 = ____
16. 222 − 89 = ____
17. 661 − 25 = ____
18. 987 − 79 = ____
19. 947 − 58 = ____
20. 901 − 38 = ____

> 'There has been no more revolutionary contribution than the one which the Indians made when they invented zero.'
>
> —Lancelot Hogben, English mathematician

Subtractions Using 'All from 9 and Last from 10'

By the way, did you know that the concept of the 'zero' was invented in India? And that the Hindu–Arabic numeral system is the commonest system for symbolic representation of numbers in the world? It was invented by Indian mathematicians between the first and fourth centuries. Isn't it then evident that Indians may have also known a thing or two about calculations too?

Let me then share with you one of the easiest methods of subtraction from India. It is so simple that you can summarize it in a short phrase known as a 'sutra'. The new method or the 'sutra' for subtraction is called **All from 9 and last from 10**. This is particularly useful when subtracting any number from 10, 100, 1000, 10,000 etc.

Say, you have to solve 1000 – 283.

In this system, you solve left to right and not right to left and then apply the maths sutra for subtraction.

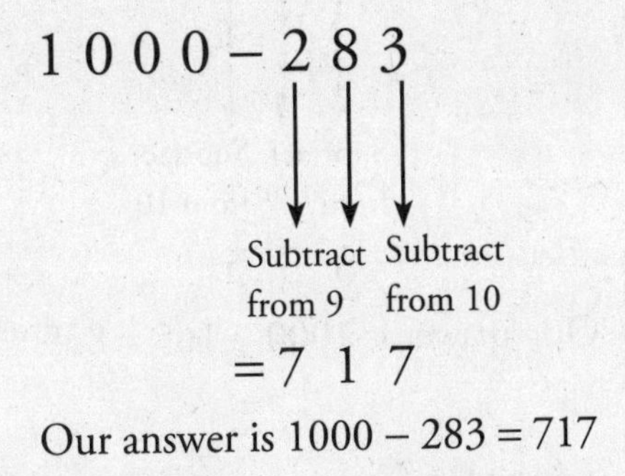

Our answer is 1000 – 283 = 717

Step 1

The maths sutra is 'All from 9 and last from 10', which means that when subtracting you will subtract all the numbers from 9 but the last one from 10. So, first you subtract the first digit on the left, which is 2, from 9. So, you have 9 – 2 = 7. This is the first digit of the answer.

Step 2

In the second step, you subtract 8 from 9, so you have 9 – 8 = 1.

Step 3

In the final step, you subtract the last digit from 10. So, you subtract 3 from 10. You have 10 – 3 = 7, which is the final digit of the answer.

So, the answer is 717.

Let's try another next example. Say, 1000 – 165

Let's apply the maths sutra 'All from 9 and last from 10'.

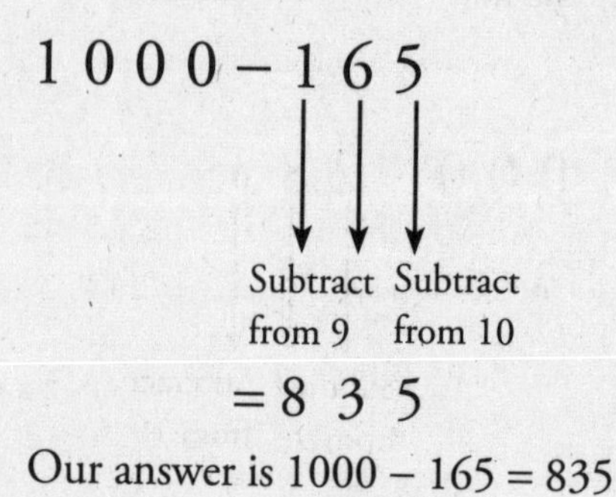

= 8 3 5

Our answer is 1000 – 165 = 835

Step 1

Here, we start from left to right and subtract 1 from 9. We get 9 – 1 = 8.

Step 2

We then subtract the middle digit 6 from 9. We have 9 – 6 = 3.

Step 3

The last digit on the right is subtracted from 10. We get 10 – 5 = 5.

The answer is 835.

This maths sutra applies to all powers of 10, like 10, 100, 1000, 10,000, 100,000 and so on. This is an easy way to subtract from base numbers like these.

Let's try to solve 10,000 – 5678 by applying the same maths sutra.

Step 1
Again, starting from left to right, we get:
9 – 5 = 4
9 – 6 = 3
9 – 7 = 2

Step 2
10 – 8 = 2
So, the final answer is 4322.

The Super Subtraction Method

Once when I was teaching this method in a class in Kolkata, a student wondered and asked me 'What happens if the number is not a base number? How do we then subtract? Is there a method for that?'

And sure enough, there was. It was called 'super subtraction'.

The 'super subtraction' method teaches us how to subtract any number from any number.

Say, we have 623 – 348. The sum must be first written in the normal way.

$$\begin{array}{r} 623 \\ -\ 348 \\ \hline \end{array}$$

Step 1

We will then subtract from left to right. We first look at the column on the left. We have 6 – 3 = 3.

$$\begin{array}{r} 623 \\ -\ 348 \\ \hline 3 \end{array}$$

Step 2

Now we move to the second step—to the next column. But here we see that 4 at the bottom is higher than 2 at the top. The result will be a negative number. So we go back to the answer row of the first column and take 1 away from 3. We now have 2 in the answer row of the first column. And we carry over 1 to the next step.

We take the 1 to the top row of the second column and prefix it to the existing digit, which is 2. So now we have 12 as the number in the top row of the second column.

12 – 4 = 8. This is the middle digit.

Our sum looks like this:

$$\begin{array}{rrrr} & & {\scriptstyle 1} & \\ & 6 & 2 & 3 \\ - & 3 & 4 & 8 \\ \hline & \not{3} & 8 & \\ & 2 & & \end{array}$$

Step 3

In the final step, we see that in the third column, 8 below is higher than 3 on the top. So, we repeat what we did in the second step.

Since there will be a negative in the final column, we go back one column—to the middle column—and reduce 8 in the answer row to 7 and carry over 1 to the digit in the top row in the last column, which is 3. This makes it 13 – 8 = 5.

$$\begin{array}{r} {\scriptstyle 11} \\ 623 \\ -\ 348 \\ \hline \not{3}\not{8}5 \\ 27 \end{array}$$

The answer is 275.

I want to add a point here. It could happen that in the first instance of solving a subtraction problem via this method, you find it difficult and may even compare it with the traditional system taught to you at school.

I felt the same way too when I was first introduced to this method. But within a day or two I could do my subtraction problems mentally, quickly, and accurately via this method. Trust me, give it another shot and I am sure you will begin to like and adore this method.

So, let's solve another subtraction problem. Let's take 425 – 247.

$$\begin{array}{r} 425 \\ -\ 247 \\ \hline \end{array}$$

Step 1

Here, we again subtract from left to right. So, we first do 4 – 2 = 2.

$$\begin{array}{r} 425 \\ -\ 247 \\ \hline 2 \end{array}$$

Step 2

In the next step, we see that there is a negative number. So, we go back one step and reduce 2 to 1 and carry over 1 to the top row of the next column—2—making it 12.

Now, we have 12 – 4 = 8. So, the middle digit becomes 8.

$$\begin{array}{r} {}^{1} \\ 425 \\ -\ 247 \\ \hline \not{2} \\ 18 \end{array}$$

Step 3

Again, in the final step, there would be a negative number. So, we go back one step and reduce 8 to 7 and carry over 1 to the top row of the third column—5—making it 15. So, we get 15 – 7 = 8. So, the answer is 178.

```
  11
  425
– 247
-----
  2̸
  18̸
   78
```

The left to right subtraction is very quick, when done mentally. You can just call out the answers at a much faster pace. Initially, some of you may take time to do it the new way, but that's okay for now. One of my friends subtracted a 70-digit number from a 70-digit number mentally, from left to right. Any guesses on the method? Yes, it was the 'super subtraction' method. It will take time, but it will show results, trust me!

ACTIVITY 2

Find the difference.

1.	**2.**	**3.**	**4.**
861	894	306	509
– 507	– 445	– 160	– 429
____	____	____	____

5.	**6.**	**7.**	**8.**
562	142	882	786
– 344	– 110	– 830	– 712
____	____	____	____

9. 813 − 175

10. 560 − 126

11. 550 − 399

12. 948 − 730

13. 503 − 273

14. 475 − 294

15. 738 − 561

16. 370 − 324

17. 881 − 482

18. 597 − 393

19. 670 − 443

20. 672 − 208

Find the difference.

21. 8,285 − 1,908

22. 7,628 − 5,678

23. 7,593 − 4,631

24. 9,065 − 3,201

25. 3,462 − 2,892

26. 8,282 − 3,189

27. 7,207 − 2,558

28. 7,193 − 3,739

29. 9,120 − 7,213

30. 2,404 − 1,701

31. 9,479 − 6,205

32. 9,507 − 1,870

33. 3,711 − 2,953

34. 4,199 − 2,879

35. 7,059 − 5,063

36. 9,553 − 6,573

37. 5,858 − 4,303

38. 6,626 − 2,250

39. 5,244 − 3,201

40. 9,369 − 6,112

As I was solving the Japanese puzzle Kakuro on the Internet for improving my own puzzle-solving and mathematical ability skills, I came across an amazing puzzle called KenKen® and since then, I have been completely captivated by it. KenKen is again from Japan and the word 'ken' means wisdom in Japanese. It was developed in 2004 by Japanese maths trainer Tetsuya Miyamoto, who wanted to make maths easier and fun for his students in various grades. Mr Miyamoto's educational philosophy boils down to 'the art of teaching without teaching'. KenKen is a registered trademark of Nextoy, LLC.

As you will see in the next few pages, KenKen is a grid-based numerical puzzle like Sudoku and is very addictive. It enhances your skills in addition, subtraction, multiplication and even division. In KenKen, you will have to think ahead. And thinking ahead is at the core of each KenKen puzzle. You will improve your chances of solving a KenKen puzzle if you look at all the events which can occur in the puzzle. So, let's take an example now and try to solve a KenKen.

Here's our first KenKen puzzle with three columns and three rows.

3+	5+	1+
		5+
4+		

So here in this 3 × 3 KenKen, we have to fill it up with figures 1, 2 or 3 without repeating the numbers in any row or column.

If you notice there are some boxes within the puzzle which have darker outlines than the others. They are called 'cages'. Also, there is a number written with a maths symbol next to it. This number can be called our 'target number' and the symbol of operation defines the rule we have to use in the box to reach the target number. In this KenKen puzzle, note that you are only using addition as the operation.

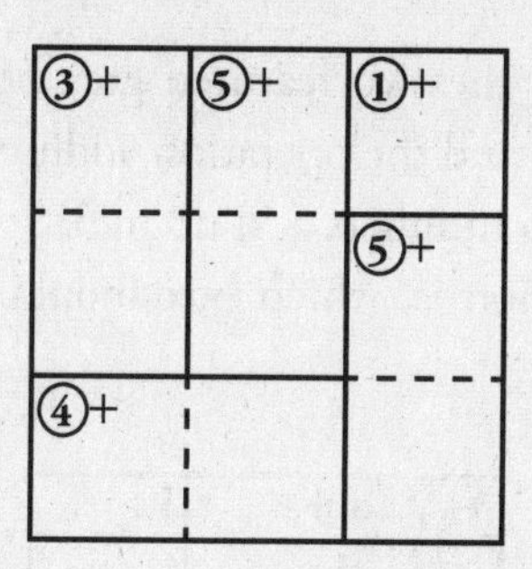

Also, note that certain boxes have only one square with one number and its operation only. This means that, the box will have that number only.

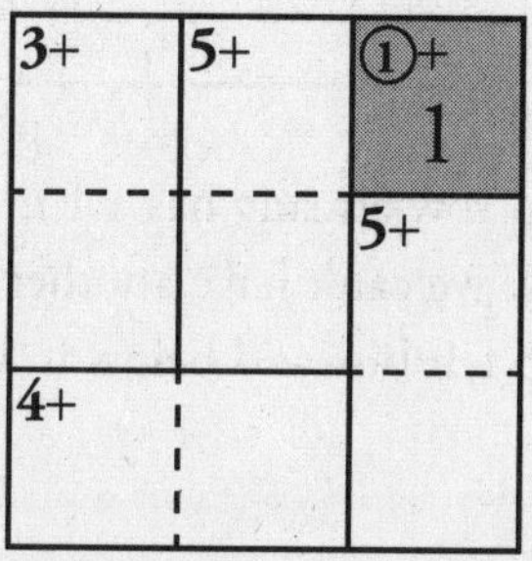

3+ 5+ 1+
1
5+
4+

In this puzzle, the dark top right corner has a single digit with one operation only. So, we will put 1 in that box.

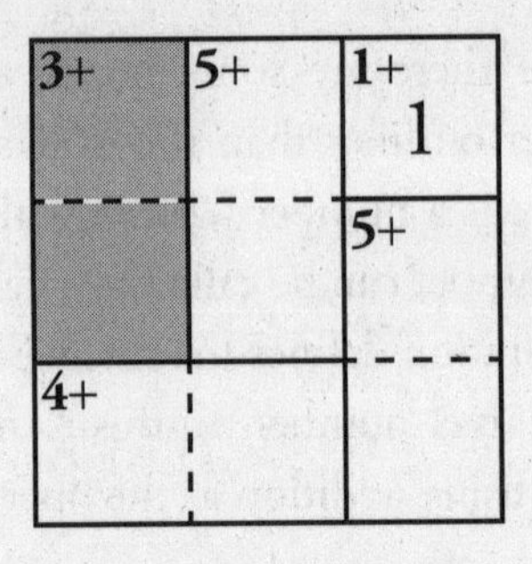

Now, do you see the two featured grey boxes above which has the number 3 and the operation addition? Now, can you tell me which two numbers add up to 3? It's got to be 1 and 2—but the question is, which box should get 1 and which box should get 2?

Notice that the top row already has a 1 in the top right box. So, in the same row we can't have another 1. So instead of 1, we put 2 in the top left box and below it, we put 1.

3+ 2	5+	1+ 1
1		5+
4+ 3		

In KenKen 3 × 3, we can only use digits from 1 to 3. So, if you notice in the first column we already have the digits 2 and 1. The only remaining digit is 3, which we put below 1—as shown above.

3+ 2	5+	1+ 1
1		5+
4+ 3	1	

So, we can now complete the 4+ box. Since we have put 3, we need 1 more. We put 1 and that completes the cage.

3+ 2	5+	1+ 1
1		5+
4+ 3	1	2

Now, the bottom row already has 3 and 1. We need one more digit which is 2. So, we put 2 in the bottom right box which completes the bottom row.

3+ 2	5+	1+ 1
1		5+ 3
4+ 3	1	2

The last column already has 1 and 2. We need 3 to be put in the middle-featured box. So, we put 3 there and that completes the last column. Also, note that putting 3 completes the cage and operation of 5+. Because, we have 3 + 2 = 5.

Now in the middle row we already have 1 and 3. We need one more digit, which can only be 2. We put 2 in the middle box, which completes the middle row. There is one more box which is empty in the first row. Using the same logic, we put 3 there, which now completes our KenKen puzzle.

3+ 2	5+ 3	1+ 1
1	2	5+ 3
4+ 3	1	2

Here is our completed KenKen. Remember, each puzzle will always have one unique solution and there can be multiple ways to reach that.

3+ 2	5+ 3	1+ 1
1	2	5+ 3
4+ 3	1	2

So, let me now share with you a few more 3 × 3 KenKen puzzles to solve.

Now note, in some of the KenKen puzzles you will see the subtraction sign. For example in the puzzle number (iii) below you will see 2–(minus). Here the goal is to make 2 using subtraction. Now you can do that using the pair (3,1) as 3 – 1 = 2. You can also use (1, 3). Please note that the order doesnt matter here.

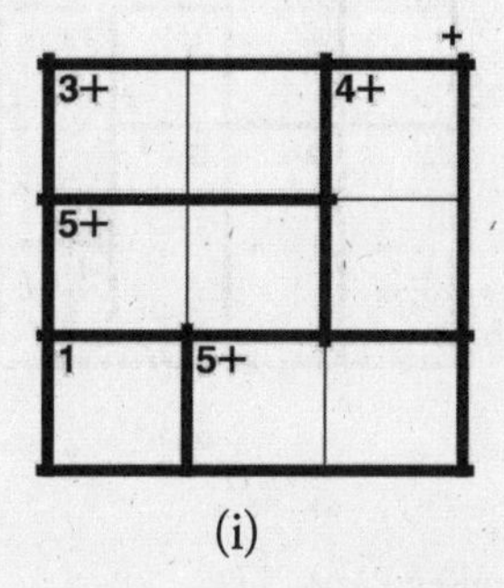

(i)

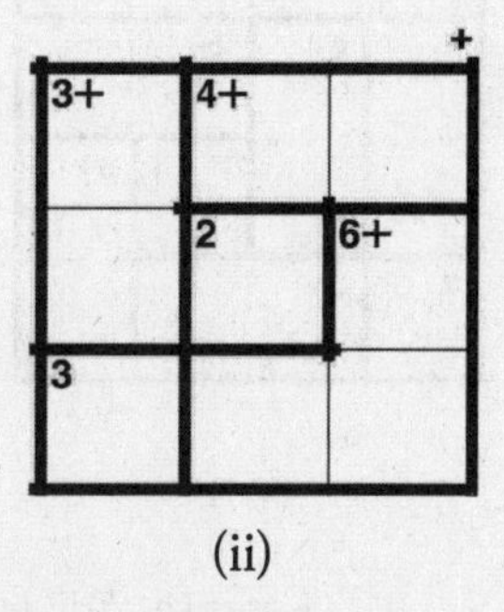

(ii)

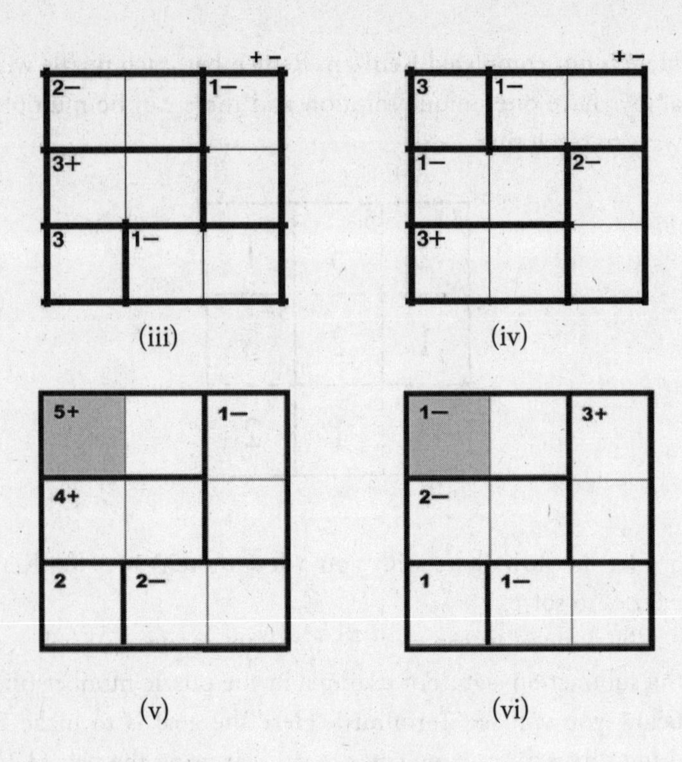

(iii) (iv)

(v) (vi)

Now, let us try two 4 × 4 KenKen Puzzles

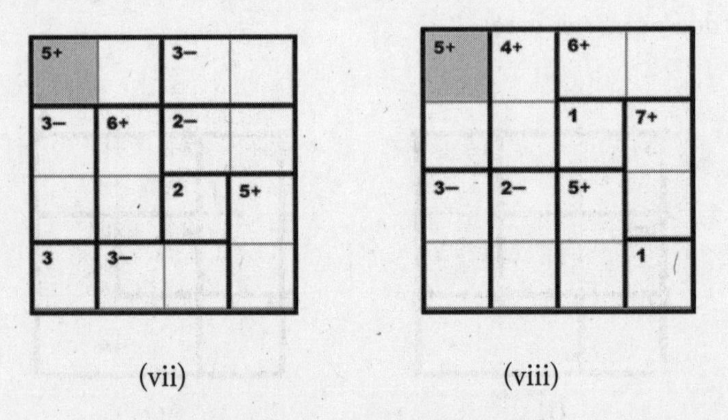

(vii) (viii)

I hope you liked KenKen. It is fun and anyone from any age can do it. If you wish to solve more puzzles of various difficulty levels and more operations, I would urge you to visit www.kenken.com. In KenKen, you can even do puzzles for operations like multiplication and division as we shall see over the next few chapters.

3

Speed up Your 'Multiplication'

Now, let me ask you a question. Think for a minute and then answer. How much time would you take to solve 98 × 97 mentally? I always ask this question at my workshops for students of all age groups and sometimes teachers too. I get answers like 4–5 minutes and some adults say a couple of minutes.

This question 98 × 97 can be solved in less than five seconds, using a principle that originated in India—the land which gave the world the concept of 'zero'. This may be a concept you may have never seen before. So, brace yourselves as I cover the method behind the solution of this question.

This method is called the 'Base method of multiplication' and this method can be used when the numbers are very close to a base like 10, 100, 1000, 10,000 and so on. So, let's look at the steps and rules in a very basic problem first.

Say, we have to multiply 8 times 7.

We note that both 8 and 7 are near to 10.

Step 1

Place the numbers with the multiplication sign as shown below and draw the lines as shown in the diagram.

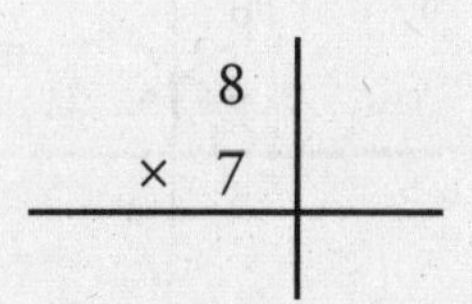

Since 8 is 2 less than 10, we place the digit 2 in the right column with a minus sign in front of it. Since 7 is 3 less than 10, we place the digit 3 in the right column with a minus sign in front of it.

8	– 2
× 7	– 3

Step 2

Now you need to remember two rules.

Rule 1: You add or subtract crosswise, as the sign (either + or –) suggests.

So, here you subtract 3 from 8, (8 – 3) or 2 from 7, (7 – 2). You could use either diagonal. You will always get the same answer.

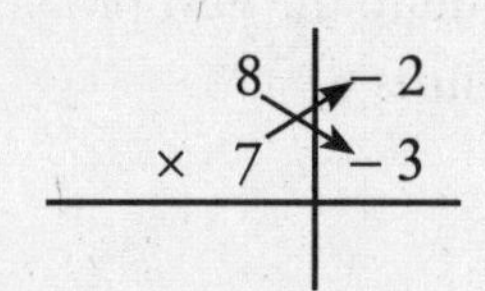

Step 3

The answer is 5. This is the first part of your answer. Write it on the left-hand side, under the line.

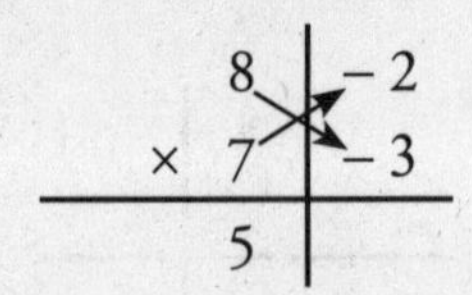

To get the second part of your answer, which is the right-hand column, you have to apply the second rule which is:

Rule 2: Multiply vertically.

Here you multiply – 2 and – 3, which gives you + 6. So, you now have the second part of the answer. The complete answer is 56.

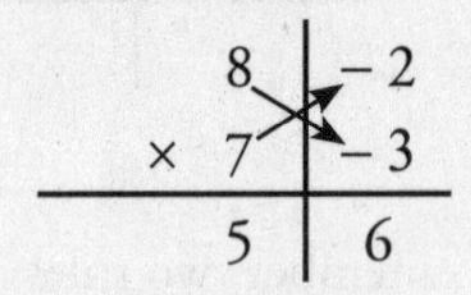

This method applies to all cases when numbers are close to 10, 100 and 1000 and so on.

I would recommend you to go over this problem one more time to make sure that you have understood it. If it's clear, you can move on to the next problem now.

Let's now try 7 times 6.

Step 1

Just like the last problem, you can place 7 and 6 above and below, with the multiplication sign.

$$\begin{array}{r|l} 7 & \\ \times\ 6 & \\ \hline & \end{array}$$

Since 7 is 3 less than 10, we place the digit 3 in the right column with a minus sign in front of it. Since 6 is 4 less than 10, we place the digit 4 in the right column with a minus sign in front of it.

Now our sum looks like this:

$$\begin{array}{r|l} 7 & -3 \\ \times\ 6 & -4 \\ \hline & \end{array}$$

Step 2

Let's now apply the rules.

Rule 1: You must add or subtract crosswise, as the sign suggests. So, we subtract 4 from 7 (7 – 4) or 3 from 6 (6 – 3). Either way, the answer is 3.

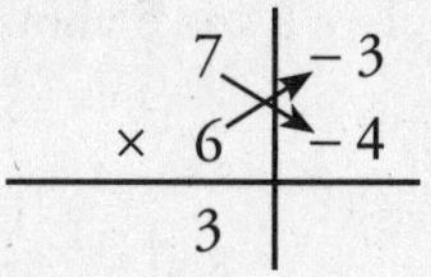

$$\begin{array}{r|l} 7 & -3 \\ \times\ 6 & -4 \\ \hline 3 & \end{array}$$

Step 3

We write 3 under the line in the left column.

Rule 2:

Now we apply the second rule—'multiply vertically'. We now multiply – 3 and – 4, which give us the answer + 12.

We now use a rule called the 'placement rule'. Since the base ten has one zero, there can be only one digit on the right-hand side. Any more digits will have to be taken to the left-hand side and added.

So, the 2 remains in the right column and the 1 is taken to the left column and added to the 3, which was already there. The sum looks like this:

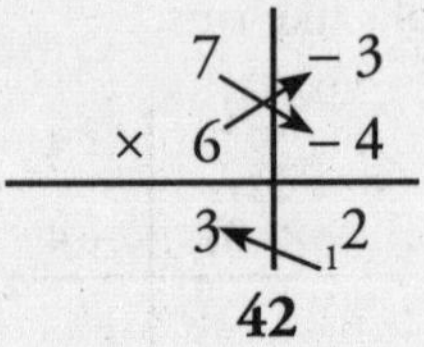

$$\begin{array}{r|l} 7 & -3 \\ \times\ 6 & -4 \\ \hline 3 & {}_{1}2 \end{array}$$

$$\mathbf{42}$$

The answer is 42.

Let's now try another one, 8 times 5.

Step 1

We draw a cross and place 8 and 5 above and below, with the multiplication sign.

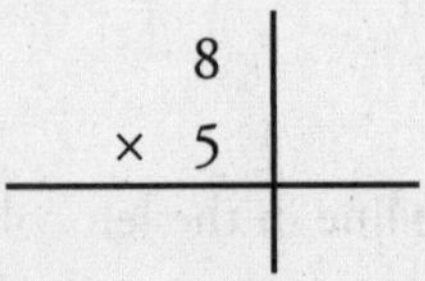

$$\begin{array}{r|l} 8 & \\ \times\ 5 & \\ \hline & \end{array}$$

Since 8 is 2 less than 10, we place the digit 2 in the right column with a minus sign in front of it. Since 5 is 5 less than 10, we place the digit 5 in the right column with a minus sign in front of it.

Now our sum looks like this:

$$\begin{array}{r|l} 8 & -2 \\ \times\ 5 & -5 \\ \hline & \end{array}$$

Step 2

We now subtract 5 from 8 (8 – 5) and 2 from 5 (5 – 2). The answer is 3 in either case. We place it on the left-hand side.

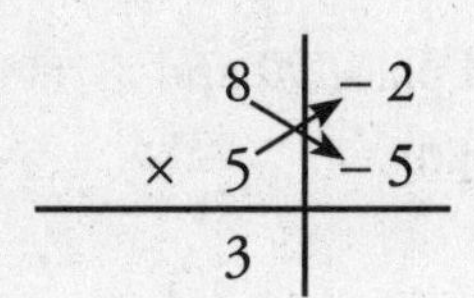

Step 3

We write 3 below the line in the left column.

We then multiply vertically – 2 and – 5 which give us the answer + 10.

We now use the placement rule. Since ten has one zero, there can be only one digit on the right-hand side. Any more digits have to be taken to the left-hand side and added.

So, the zero remains in the right column and the 1 is taken to the left column and added to the 3, which was already there. Now, the sum looks like this:

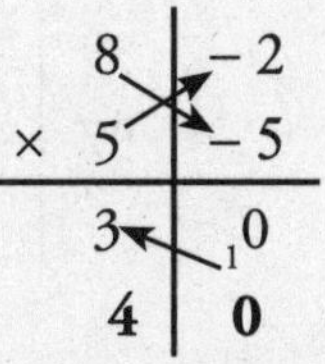

$$\begin{array}{r|l} 8 & -2 \\ \times\ 5 & -5 \\ \hline 3 & {}_{1}0 \\ \mathbf{4} & \mathbf{0} \end{array}$$

So, our answer is 40.

So, do you like this new base method so far? I hope you do and remember, we are just getting started now with the bigger problems.

So far we have tried multiplying numbers close to 10. Now, let's use this method to multiply numbers closer to 100. Let's now calculate 99 × 97. And let's do it in five seconds!

Let's take 99 × 97 as an example.

Step 1

We draw a cross and place 99 and 97 above and below, with the multiplication sign.

$$\begin{array}{r|l} 99 & \\ \times\ 97 & \\ \hline & \end{array}$$

Since 99 and 97 are closer to 100 than to 10, we take 100 as the base. Now 99 is less than 100 by 1. So, we write – 01 in the right-hand column (top). The number 97 is less than 100 by 3. So, we write – 03 in the right-hand column (bottom). We write – 01 and – 03 and not – 1 and – 3, because the base is 100 and since 100 has two zeros on the right-hand side, we know there has to be two digits. So, now our sum looks like this:

$$\begin{array}{r|l} 99 & -01 \\ \times\ 97 & -03 \\ \hline & \end{array}$$

Step 2

Let's apply the first rule, which is to add or subtract crosswise. So, we calculate 99 – 03 = 96. Also note that if we had done 97 – 01, we would get the same answer, 96. So, we write 96 under the line in the left-hand column.

99	– 01
× 97	– 03
96	

Step 3

To arrive at the final answer, we now apply the second rule, which is to multiply the numbers in the right column vertically. We multiply – 01 with – 03 and get + 03. We now write this under the line in the right-hand column.

99	– 01
× 97	– 03
96	03

Our complete answer is 9603.

Let's now come to the example you've perhaps been waiting for since the beginning of this chapter. 'Can you multiply 98 × 97 in less than five seconds?'

98	
× 97	

Step 1

Now, look at the first step. Since 98 and 97 are close to 100, we take 100 as the base. The number 98 is 2 less than 100, so we write – 02 in the right-hand column (top). And similarly, for 97, we write – 03. We have written – 02 and – 03 and not – 2 and – 3, because the base is 100. Since 100 has two zeroes, there has to be two digits on the right-hand side. So, now our sum will look like this:

98	– 02
× 97	– 03

Step 2

We now apply rule 1, which is, you add or subtract crosswise, as the sign (either + or –) suggests. So, we have 98 – 03 = 95. If we take the other diagonal, 97 – 02, we will get the same answer, which is 95. We write that below the line in the left-hand column. So, our sum now looks like this:

98	– 02
× 97	– 03
95	

Isn't this interesting?

In our final step, we will multiply the numbers in the right-hand column vertically. So, we have – 02 × – 03, giving us + 06. We write that down below the line in the right-hand column and we have the answer, which is 9506. Now, how much time did that take? Captivating, isn't it?

$$\begin{array}{r|l} 98 & -02 \\ \times\ 97 & -03 \\ \hline 95 & 06 \end{array}$$

Now, do you know who discovered these quick methods?

These methods were discovered by an Indian saint, Tirthaji, in the early twentieth century. Tirthaji had a brilliant academic record with a master's degree in six subjects, including English, Sanskrit and mathematics. After a brilliant university career, he became a lecturer in mathematics and science in the Baroda College in India.

As the legend goes, Tirthaji meditated for over seven years and discovered the 'mathematical sutras' from ancient Indian texts. Tirthaji then wrote a book, *Vedic Mathematics*, which was published posthumously in 1965. In the path-breaking book, Tirthaji gave word formulas, also known as 'sutras', with which you could calculate seemingly difficult maths problems, mentally!

If you would ask Tirthaji whether this was maths or magic? He would reply, 'It is both. It is magic until you understand it and it is mathematics thereafter.'

Let's try 88 times 88.

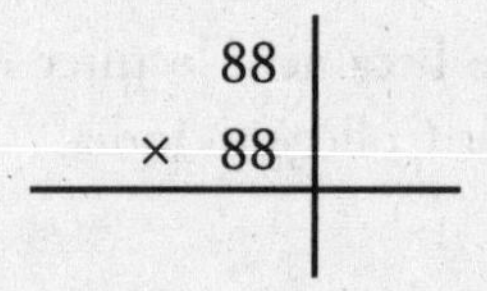

Step 1

Since 88 is close to 100, we take 100 as the base. The number 88 is 12 less than 100 so we write –12 in both rows of the right-hand column. The number 88 – 12 is 76, so we write 76 below the line in the left-hand column.

88	–12
× 88	–12
76	

Step 2

Then we multiply the figures in the right-hand column vertically. – 12 × – 12 = 144. We write 144 below the line in the right-hand column.

$$\begin{array}{r|l} 88 & -12 \\ \times\ 88 & -12 \\ \hline 76 & \end{array}$$

Step 3

Now, the right-hand column had three digits, but according to the placement rule, since the base is 100, and 100 has two zeroes, on the right-hand side, there can be two digits only. So, the extra digit should be carried forward to the left. So, the 1 from 144 is moved to the left-hand column and added to 76. Our sum looks like this now:

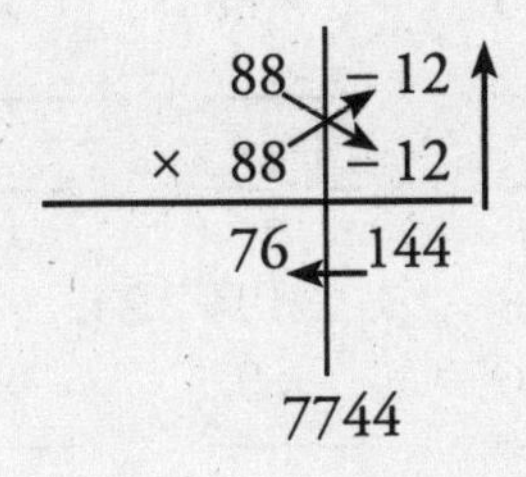

The answer is 7744.

ACTIVITY 1

Find the product.

1. 88 × 98

2. 87 × 96

3. 92 × 89

4. 96 × 95

5. 88 × 91

6. 96 × 89

7. 95 × 98

8. 96 × 94

9. 90 × 98

10. 94 × 99

11. 96 × 92

12. 94 × 96

13. 96 × 90

14. 86 × 95

15. 88 × 96

16. 99 × 97

17. 88 × 90

18. 95 × 88

19. 94 × 90

20. 92 × 92

21. 97 × 94

22. 98 × 96

23. 94 × 92

24. 89 × 95

25. 88 × 99

26. 93 × 96

27. 89 × 94

28. 97 × 97

29. 97 × 90

30. 98 × 99

31. 94 × 89 ______

32. 91 × 94 ______

33. 89 × 92 ______

34. 94 × 88 ______

35. 93 × 88 ______

36. 95 × 92 ______

37. 87 × 93 ______

38. 88 × 93 ______

39. 90 × 96 ______

40. 86 × 90 ______

41. 97 × 92 ______

42. 94 × 93 ______

43. 93 × 97 ______

44. 87 × 94 ______

45. 99 × 89 ______

46. 97 × 88 ______

47. 94 × 91 ______

48. 92 × 97 ______

49. 92 × 91 ______

50. 92 × 88 ______

You can also note that the same rules and principles apply to higher bases as well, like base 1000, 10,000 etc.

Let's take an example: 998 × 996

$$\begin{array}{r|l} 998 & \\ \times\ 996 & \\ \hline & \end{array}$$

Step 1

Both the numbers are close to 1000, so we take 1000 as the base number. The number 998 is less than 1000 by 2, so we write – 002 in the right-hand column (top). Since 996 is less than 1000 by 4, we write – 004 in the right-hand column (bottom). Now our sum looks like this:

$$\begin{array}{r|l} 998 & -002 \\ \times\ 996 & -004 \\ \hline & \end{array}$$

Step 2

As we had done before, we subtract crosswise. So, we have, 998 – 004 = 994. So, we write 994 below the line in the left-hand column. If we had done 996 – 002, we would get the same answer, which is 994.

$$\begin{array}{r|l} 998 & -002 \\ \times\ 996 & -004 \\ \hline 994 & \end{array}$$

Step 3

We now multiply – 002 with – 004. The answer is + 008. We write 008 below the line in the right-hand column. Our answer is 994008.

$$\begin{array}{r|l} 998 & -002 \\ \times\ 996 & -004 \\ \hline 994 & 008 \end{array}$$

Let's take another similar sum: 989 × 934.

989	
× 934	

Step 1

Since both the numbers are close to 1000, we take 1000 as the base number. The number 989 is less than 1000 by 11, so we write – 011 in the right-hand column (top). The number 934 is less than 1000 by 66, so we write – 066 in the right-hand column (bottom). The sum looks like this now:

989	– 011
× 934	– 066

Step 2

We now subtract crosswise. So, we have, 989 – 066 = 923, so we write 923 below the line in the left-hand column.

989	– 011
× 934	– 066
923	

Step 3

We then multiply the numbers in the right-hand column vertically. So, it will be – 011 × – 066 = 726. So, we write 726 in the right-hand column.

	998	– 011
×	934	– 066
	923	726

Our answer becomes 923726.

Let's take a similar sum with the carry-over step. So, we have: 985 × 926.

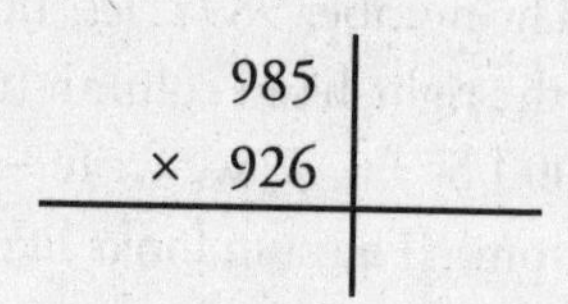

Step 1

The first step is the same as before. Since both numbers are close to 1000, we take 1000 as the base number. The number 985 is less than 1000 by 15, so we write – 015 in the right-hand column (top). The number 926 is less than 1000 by 74, so we write – 074 in the right-hand column (bottom). The sum now looks like this:

	985	– 015
×	926	– 074

Step 2

We now subtract crosswise: 985 – 074 = 911. So, we write 911 below the line in the left-hand column. If we do 926 – 015, we will get the same answer, which is 911.

$$\begin{array}{r|l} 985 & -015 \\ \times\ 926 & -074 \\ \hline 911 & \end{array}$$

Step 3

We now multiply the numbers in the right-hand column vertically: – 15 × – 74 = 1110.

We then apply the placement rule. Since thousand has three zeroes, there should be only three digits on the right-hand side. Any extra digits must be carried forward and added to the left-hand column. So, we add 1, which is the extra digit, to 911 to get 912.

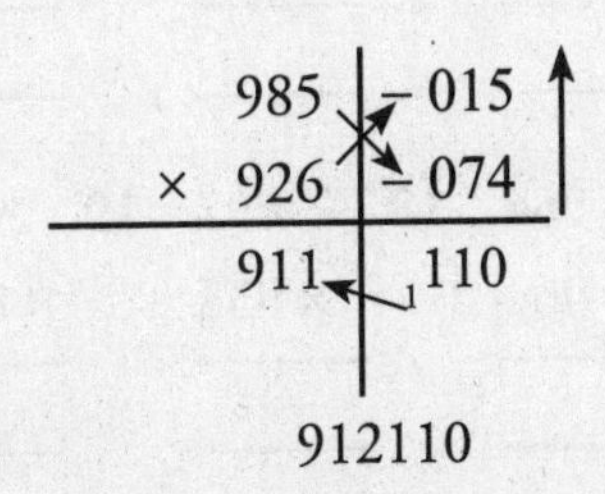

So, our answer is 912110.

ACTIVITY 2

1. 942 × 998	**2.** 990 × 889	**3.** 984 × 905	**4.** 938 × 916	**5.** 923 × 934
6. 924 × 985	**7.** 958 × 898	**8.** 909 × 958	**9.** 982 × 945	**10.** 957 × 993
11. 896 × 975	**12.** 938 × 990	**13.** 942 × 906	**14.** 980 × 950	**15.** 972 × 903
16. 969 × 895	**17.** 964 × 964	**18.** 897 × 938	**19.** 964 × 891	**20.** 973 × 899
21. 897 × 967	**22.** 952 × 900	**23.** 983 × 957	**24.** 982 × 947	**25.** 931 × 964
26. 959 × 979	**27.** 967 × 916	**28.** 900 × 907	**29.** 927 × 982	**30.** 898 × 951

31. 900 × 909	**32.** 915 × 943	**33.** 957 × 944	**34.** 980 × 962	**35.** 929 × 924
36. 938 × 957	**37.** 933 × 979	**38.** 907 × 908	**39.** 896 × 903	**40.** 965 × 930
41. 924 × 896	**42.** 906 × 934	**43.** 907 × 909	**44.** 966 × 948	**45.** 926 × 900
46. 938 × 944	**47.** 952 × 896	**48.** 968 × 944	**49.** 968 × 936	**50.** 928 × 891

When I introduce this to students in the fourth or fifth standards, someone always gets up and asks me this question 'Sir, the problems which you showed us are below the base 10 like 9, 8 or below the base 100, like 99, 98 or 97. Can there be problems which are above the base?'

So, I always respond, 'Let's experiment and find out for ourselves. Let's take an example say 12 × 13.'

$$\begin{array}{r|l} 12 & \\ \times\ 13 & \\ \hline & \end{array}$$

Step 1

Earlier, the numbers we had seen were below 10, here we see that both 12 and 13 are more than 10. Since 12 is more than 10 by 2, we write +2 in the right-hand column (top). Similarly, for 13, we write +3 in the right-hand column (bottom).

12	+ 2
× 13	+ 3

Step 2

We now add across the diagonal. Remember, here we add, following the plus sign. Adding crosswise we get 12 + 3 = 15. If we add 13 + 2, we will get the same answer. So, 15 is the first part of our answer, which we write below the line in the left-hand column.

12	+ 2
× 13	+ 3
15	

Step 3

Our final step is to multiply vertically the numbers in the right-hand column. So (+ 2) × (+ 3) = + 6. We write 6 below the line in the right-hand

12	+ 2
× 13	+ 3
15	6

Our answer is 156.

Can we now try 14 × 13?

$$\begin{array}{r|l} 14 & \\ \times\ 13 & \\ \hline & \end{array}$$

Step 1

Here again the base is 10, as both 14 and 13 are close to 10. Since 14 is more than 10 by 4, we write + 4 in the right-hand column (top). Similarly, for 13, we write +3 in right-hand column (bottom).

$$\begin{array}{r|l} 14 & +4 \\ \times\ 13 & +3 \\ \hline & \end{array}$$

Step 2

Now, once again we add across the diagonal. We add because there is a plus sign. Adding crosswise, we get: 14 + 3 = 17. If we add 13 + 4, we will get the same answer. So, 17 is the first part of our answer, which we write below the line in the left-hand column.

$$\begin{array}{r|l} 14 & +4 \\ \times\ 13 & +3 \\ \hline 17 & \end{array}$$

Step 3

In our final step, we will vertically multiply the numbers in the right-hand column. Multiplying 4 by 3 gives us 12. Now our base is 10. So, we apply the placement rule. Since 10 has one zero, there will be only one digit on the right-hand side. The number 12 has two digits, so we carry over 1 to the left-hand side and add it to 17. So, we have: 17 + 1 = 18. You write 18 under the line in the left-hand column.

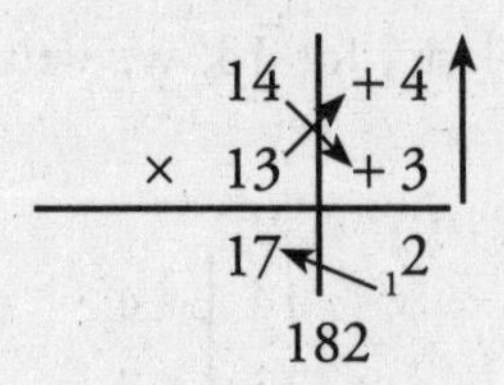

The answer is 182. Isn't this so much fun?

Let's take a new problem of the same nature. Consider the following multiplication: 15 × 16.

15
× 16

Step 1

Just as before, we will write the excesses from our base 10 on the right-hand side. The number 15 is 5 more than 10, so we write + 5 in the right-hand column (top) and similarly for 16, we write + 6 in the right-hand column (bottom).

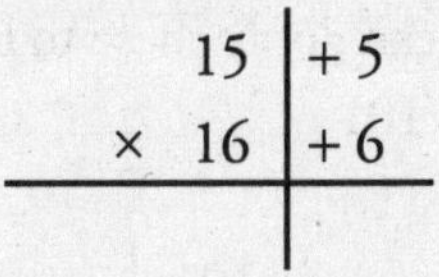

$$\begin{array}{r|l} 15 & +5 \\ \times\ 16 & +6 \\ \hline & \end{array}$$

Step 2

Now, we add along the diagonal. So, we have 15 + 6 = 21 or 16 + 5 = 21.

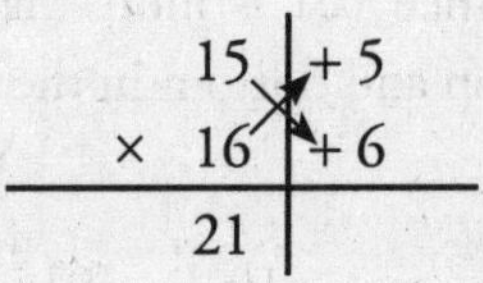

$$\begin{array}{r|l} 15 & +5 \\ \times\ 16 & +6 \\ \hline 21 & \end{array}$$

Step 3

In our final step, we multiply the numbers in the right-hand column vertically. So, we have 5 × 6 = 30. Since our base is 10, we apply the placement rule. The number 10 has one 0, so there will be only one digit on the right-hand side. The number 30 has two digits, so we carry over 3 to the left-hand side and add it to 21. So, now we have, 21 + 3 = 24. You write 24 under the line in the left-hand column.

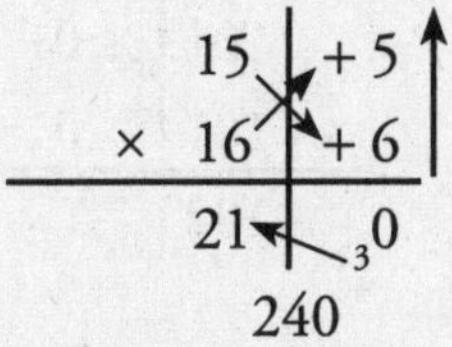

$$\begin{array}{r|l} 15 & +5 \\ \times\ 16 & +6 \\ \hline 21 & {}_{3}0 \end{array}$$

$$240$$

The answer is 240.

Let us now see if we can apply these to base 100.
Let's calculate 101 × 101.

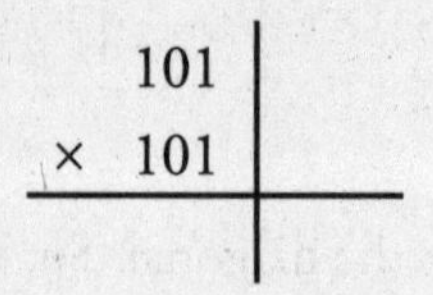

Step 1

Our base is 100. Since 101 is more than 100 by + 01, we write + 01 on the top and bottom in the right-hand column.

101	+ 01
× 101	+ 01

Step 2

Now we add crosswise. So, we have 101+01=102. This is the first part of the answer, which we write below the line in the left-hand column.

101	+ 01
× 101	+ 01
102	

Step 3

Now we will multiply the digits in the right-hand column. (+ 01) × (+ 01) = 01. We write this below the line in the right-hand column.

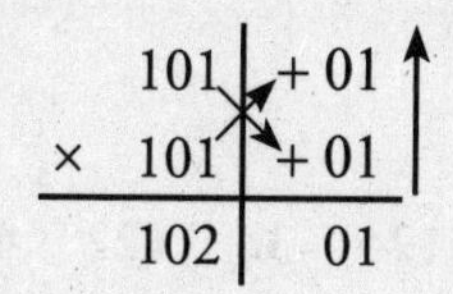

The answer is 10201.

Now try to solve the following problem. And see if you can do it mentally.

102	
× 102	

The answer is 10404.

ACTIVITY 3

1. 107 × 113	**2.** 124 × 115	**3.** 122 × 102	**4.** 104 × 116	**5.** 109 × 115
6. 116 × 105	**7.** 100 × 112	**8.** 117 × 123	**9.** 124 × 112	**10.** 105 × 103
11. 109 × 117	**12.** 104 × 111	**13.** 106 × 122	**14.** 101 × 118	**15.** 118 × 105
16. 109 × 119	**17.** 108 × 115	**18.** 116 × 118	**19.** 122 × 107	**20.** 116 × 122
21. 106 × 111	**22.** 114 × 120	**23.** 113 × 124	**24.** 114 × 115	**25.** 106 × 110
26. 121 × 115	**27.** 116 × 121	**28.** 102 × 109	**29.** 115 × 103	**30.** 114 × 123
31. 111 × 111	**32.** 100 × 103	**33.** 101 × 111	**34.** 102 × 119	**35.** 117 × 118

36.	104 × 104	**37.**	118 × 121	**38.**	108 × 112	**39.**	102 × 107	**40.**	104 × 110
41.	115 × 104	**42.**	120 × 112	**43.**	118 × 119	**44.**	108 × 125	**45.**	121 × 111
46.	125 × 104	**47.**	111 × 117	**48.**	122 × 106	**49.**	101 × 122	**50.**	115 × 106

I am sure, that now you must be wondering how we can mentally multiply any number by any number. So, let me share with you a pattern via which you can achieve this. This new method for multiplying is called the 'vertically and crosswise pattern' method. This is a formula that fits all the types of multiplication problems, so it does not matter whether or not the numbers are near the bases of 10, 100, 1000 and the like.

Let us multiply 23 × 12.

Step 1

Write the two-digit numbers one below the other with the multiplication sign, as you would for any multiplication sum.

$$\begin{array}{r|l} 23 & \\ \times\ 12 & \\ \hline & \end{array}$$

Now look at the vertically and crosswise pattern below and understand the steps.

2-Digits Vertically and Crosswise Pattern

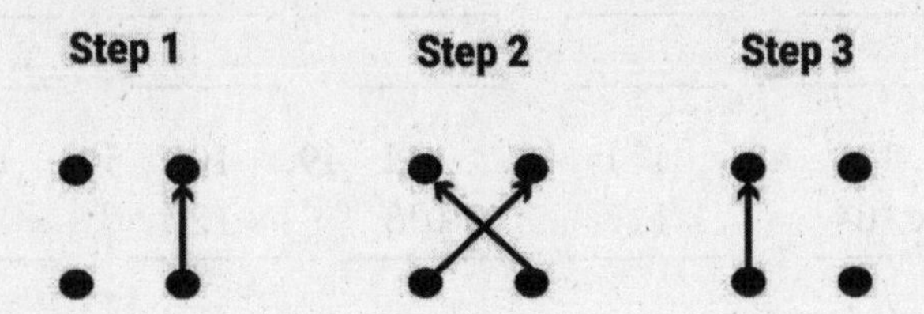

First look at the above pattern starting at **Step 1**. We start with multiplying of numbers in the unit's place. We multiply the unit's digits vertically.

So 2 × 3 = 6. We put down 6, under the line in the unit's place.

$$\begin{array}{r} 23 \\ \times\ 12 \\ \hline 6 \end{array}$$

Step 2

Now look at **Step 2** given in the pattern above. We will now do crosswise multiplication and add the numbers. We have 2 × 2 = 4 and we have 1 × 3 = 3. We take 4 and 3 and add them. So, we have 4 + 3 = 7, which will be middle digit of our answer. Our sum now looks like this.

$$\begin{array}{r} 23 \\ \times\ 12 \\ \hline 76 \end{array}$$

Step 3

In our final step, we look at the pattern above at **Step 3**. Multiply 1 × 2 = 2. This is the final digit of the answer.

$$\begin{array}{r} 23 \\ \times\ 12 \\ \hline 276 \end{array}$$

So, the complete answer is 276.

I hope this is quite clear. Let's now look at another sum which has carry-overs.

Let's try 72 × 18.

$$\begin{array}{r} 72 \\ \times\ 18 \\ \hline \end{array}$$

Step 1

Like before, we will start with the numbers in the unit's place. We multiply the unit digits, 8 × 2 = 16. Now the unit takes only one digit, so we put down 6 in the unit's place under the line and carry over the 1 to the next step.

$$\begin{array}{r} 72 \\ \times\ 18 \\ \hline {}_{1}6 \end{array}$$

Step 2

We then look at the second step and do crosswise multiplication and add the numbers. We multiply 8 × 7 = 56 and 1 × 2 = 2. We then add those two numbers: 56 + 2 = 58. We then carry over 1 from the unit's place. So, we add 1 to 58 to get 59. We put 9 down below the line and carry over 5 to the next step.

$$\begin{array}{r} 72 \\ \times\ 18 \\ \hline {}_{5}9{}_{1}6 \end{array}$$

Step 3

In the final step, we multiply vertically, 1 × 7 = 7. But there was 5 carried over from the earlier step. So, we get 5 + 7 = 12. We put down 12 under the line to get the answer 1296.

$$\begin{array}{r} 72 \\ \times\ 18 \\ \hline 1296 \end{array}$$

Now let's solve another sum which is difficult because it has a lot of carry-overs. But I am sure that if you remember

the technique of vertically and crosswise, this should be a cakewalk for you.
Let's multiply 67 × 54.

$$\begin{array}{r} 67 \\ \times\ \ 54 \\ \hline \end{array}$$

Step 1
First, we multiply 4 × 7 = 28. We put down 8 in the unit's place under the line and carry over 2 to the next step.

$$\begin{array}{r} 67 \\ \times\ \ 54 \\ \hline {}_{2}8 \end{array}$$

Step 2
In the second step, we do crosswise multiplication as before and then add the numbers. To this total, we add the carry-over amount. So, we have 5 × 7 = 35 and 4 × 6 = 24.
35 + 24 + 2 (carry-over) = 61

We put down 1 below the line and carry over 6 to the next step.

$$\begin{array}{r} 67 \\ \times\ \ 54 \\ \hline {}_{6}1{}_{2}8 \end{array}$$

Step 3

In the final step, we multiply vertically again. So, we have $5 \times 6 = 30$.

We add the carry-over amount, 6, to 30. The result is 36.

We put down 36 and have our answer, which is 3618.

$$\begin{array}{r} 67 \\ \times\ 54 \\ \hline 3618 \end{array}$$

At this stage I would urge you all to share these methods with your friends. Maths is so much fun. With this knowledge you can now breeze through the maths syllabus in school. Right?

ACTIVITY 4

1. 75×34	**2.** 96×33	**3.** 12×56	**4.** 66×50	**5.** 90×73
6. 99×94	**7.** 58×73	**8.** 97×17	**9.** 78×29	**10.** 62×80
11. 16×91	**12.** 97×44	**13.** 58×55	**14.** 38×82	**15.** 52×27
16. 24×78	**17.** 69×47	**18.** 41×60	**19.** 14×99	**20.** 19×59
21. 62×55	**22.** 61×86	**23.** 89×19	**24.** 97×15	**25.** 68×29
26. 34×79	**27.** 26×95	**28.** 66×67	**29.** 28×49	**30.** 74×43
31. 54×16	**32.** 81×77	**33.** 52×94	**34.** 87×64	**35.** 82×17

36.	**37.**	**38.**	**39.**	**40.**
21 × 56	86 × 68	40 × 22	66 × 23	45 × 85

41.	**42.**	**43.**	**44.**	**45.**
14 × 41	36 × 58	74 × 78	50 × 53	77 × 11

46.	**47.**	**48.**	**49.**	**50.**
17 × 94	14 × 54	80 × 77	13 × 78	75 × 45

All right! So, now its KenKen puzzle time again. This time we will use the multiplication and division operation to solve our KenKen puzzles. The rules are just like the last time, I hope they are all clear. These KenKen puzzles will test your skills in mathematics as well as logical thinking. And mind you, it's addictive. I am sure you will get hooked on to them. Let's get started with a 3 × 3 KenKen.

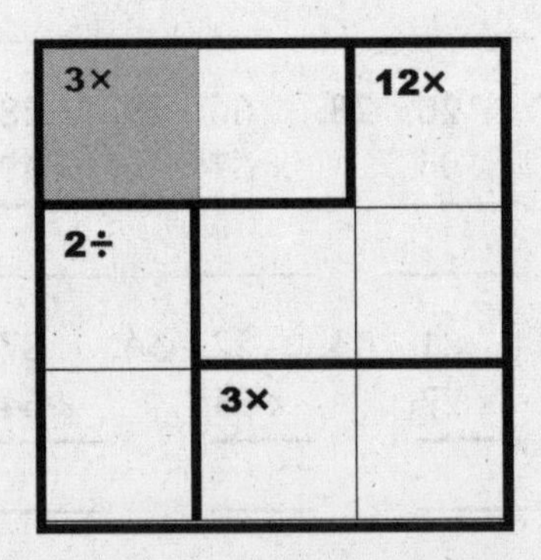

KenKen is a registered trademark of Nextoy, LLC.

Step 1

We have to fill this grid with numbers 1, 2 and 3. This puzzle can be approached from anywhere, but I will suggest approaching it from the 3 × cage on the row at the bottom. Our options to get 3 could be 3 × 1 or 1 × 3. Let's put 3 and 1 in the cage below. Our puzzle looks like this now. We have put a star against the digits 3 and 1, so we know that they are temporary and we can change them at any time if the puzzle becomes invalid.

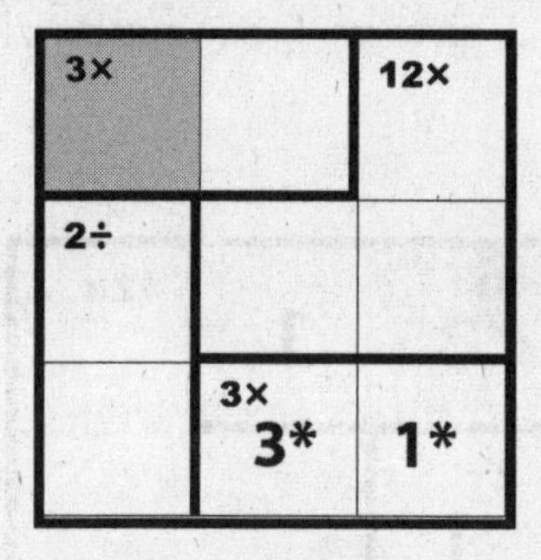

Step 2

We then move to the left of this cage and write 2, because we have already put the digits 3 and 1 in the bottom row. Now in the new cage, we can put 1 also to get 2 ÷. Our puzzle now looks like this:

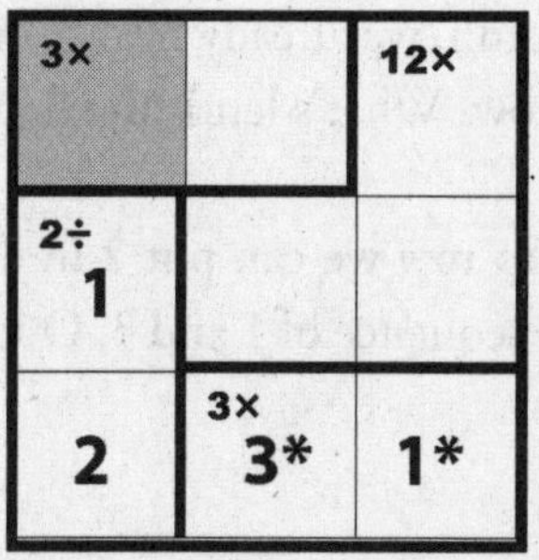

3×		12×
2÷ 1		
2	3× 3*	1*

Here when you are seeing the division sign – just like subtraction, the goal is to make 2 using division. Now you can do that using the pair (2, 1) as 2 ÷ 1 = 2. You can also use (1, 2). Please note that the order doesnt matter here in the cage.

Step 3

So now in the top 3 × across cage, we can put 3 and 1, as the product of 3 and 1 gives us 3. So, our puzzle now looks like this:

3× 3	1	12×
2÷ 1		
2	3× 3*	1*

Step 4

In the middle column, we already have 1 on the top row and 3 on the bottom row. What's left is the figure 2. We put 2 in the middle row.

Also, in the top row we can put 2 in the blank space, so as to complete the sequence of 1 and 3. Our puzzle now looks like this:

3× 3	1	12× 2
2÷ 1	2	
2	3× 3*	1*

Step 5

In the last column, we already have 1 and 2. What's left is 3. We put that in the blank space and our puzzle is complete and now looks like this.

3× 3	1	12× 2
2÷ 1	2	3
2	3× 3*	1*

So we get $12 = 2 \times 2 \times 3$ by multiplying $2 \times 2 \times 3$ in the $12 \times$ cage. Hope this is clear?

So, 'Ken-gratulations' as KenKen founder Mr Miyamoto would say! Rush now to www.kenken.com for more solving puzzles with varied difficulty levels.

4

'Divide' like an Egyptian

I just saw the new Hollywood blockbuster *The Mummy* with Tom Cruise in the main lead. And I was fascinated by the storyline set against the backdrop of the Egyptian civilization thousands of years ago. Ancient Egypt was a land of mysteries. It had mesmerized the entire planet for centuries.

Its religion and temple architecture, the Pyramids and the Sphinx—are all shrouded in a thick veil of mystery, replete with their own secrets. The Egyptian pyramids are one of the most famous of all ancient monuments, one of the wonders of the bygone world.

Egyptians used the simple concepts of addition and subtraction to multiply and divide. They did not develop formulae. The Egyptians knew that the world works in powers of 2, so they used it extensively to their advantage. They also know that any number can be expressed as the sum or difference of numbers which are all powers of 2. For example, 19. It is not a power of two but $19 = 2^4 + 2^1 + 2^0$.

Again $27 = 2^5 - 2^2 - 2^0$.

Let us first see a division problem—the Egyptian style. Say we have to divide $2116 \div 23$.

Step 1

So, we first write down the powers of 2 on our left, starting with $2^0 = 1$. We then double 1 and get 2, we double 2 and get 4, and so on; i.e., we write the successive powers of 2 like 2^1, 2^2, etc.

$2116 \div 23$

1
2
4
8
16
32
64

Step 2

We now write, in the right-hand column, our divisor 23 against 1 and go on doubling 23 successively and writing the numbers against the corresponding numbers in the left-hand column. That is, we first write 23, then double it to get 46 and then double further to get 92. The number 92 doubles to 184 and then 184 to 368 and then double 368 to 736 and finally we stop at 1472. Now, we stop at 1472 because the next double gives us 2944 which will be higher than 2116.

Note: We keep doubling the divisor in the right hand side column. In this case, we stop at 1472 because if we double this further we will get 2944 which will be higher than our dividend 2116. Once the right hand side column is achieved, we can then write down the powers of 2 in the left hand side column.

Our right-hand column looks like this:

2116 ÷ 23

1	23
2	46
4	92
8	184
16	368
32	736
64	1472

Step 3

Now the objective on the right-hand column is to find a series of numbers which add up to 2116. So, we get there by adding 1472 + 368 + 184 + 92 = 2116. We mark them on our table and refer to the terms, opposite them in the left column.

2116 ÷ 23

1		23
2		46
4	←	92
8	←	184
16	←	368
32		736
64	←	1472

So, the term opposite 1472 is 64, the term opposite 368 is 16, opposite 184 is 8 and finally the term opposite 92 is 4.

Now to arrive at our answer of 2116 ÷ 23, we simply add 64 + 16 + 8 + 4 = 92.

So, the answer of 2116 ÷ 23 is 92.

Note that now this is done by simple application of addition and subtraction. Rote learning of multiplication tables is not necessary at all. This makes division very simple and easy to do.

Let's take another example. Say, we have 1806 ÷ 7

Step 1

So, we first write down the powers of 2 on our left, starting with $2^0 = 1$. We then double 1 and get 2, we double 2 and get 4 and so on. The value of the right hand side column will be upto less than 1806 – our dividend. Here we stop at 1792 because beyond it, we will get double of 1792 which is 3584. So we need to stop at 1792 only.

1806 ÷ 7

1
2
4
8
16
32
64
128
256

Step 2

We now write down the right-hand side column. First comes 7 and then the double of 7 which is 14 and then 28 and so on, all the way up to 1792. We go up to 1792 because beyond that would be greater than our dividend 1806. Our right-hand side column looks like this:

1806	**÷ 7**
1	7
2	14
4	28
8	56
16	112
32	224
64	448
128	896
256	1792

Step 3

Now the objective of the right-hand side column is to find a series of numbers which add up to 1806. So, we get there by adding 1792 + 14 = 1806. We mark them on our table and refer to the terms, opposite these two in the left-hand side column.

1806		**÷ 7**
1		7
2	←	14
4		28
8		56
16		112
32		224
64		448
128		896
256	←	1792

So, we have 256 and 2, which we add to get our answer 256 + 2 = 258. It's as simple as that!

Now I know you have a question in your mind. You must be thinking what happens if there is a remainder? What happens if it doesn't divide out evenly? So, this is what I will show you next!

Let's take a problem, say 1401 ÷ 25. Now let's set it up like before.

Step 1

So, as usual we write down the powers of 2, all the way up to 32, in the left-hand side column.

<u>**1401 ÷ 25**</u>

1
2
4
8
16
32

Step 2

We then start doubling our divisor, which is 25. So, we get 25, double of that is 50, then double of 50 is 100, double of 100 is 200 and so on, all the way up to 800. We stop at 800 because if we double 800, we will get 1600. This will be more than 1401 and we don't want that. So, our right-hand side column looks like this:

1401 ÷ 25	
1	25
2	50
4	100
8	200
16	400
32	800

Step 3

Now the objective on the right-hand side column is to find a series of numbers which add up to 1401. So, we get there by adding 800 + 400 + 200 = 1400. But that's not 1401. So, the Egyptians had a way to deal with this.

We then subtract 1401 – 1400 = 1, which is our remainder.

Also note, we now have to consider the terms opposite 800, 400 and 200. So, we have 32 + 16 + 8 = 56

1401 ÷ 25	
1	25
2	50
4	100
8 ⟵	200
16 ⟵	400
32 ⟵	800

So, our answer looks like $56\frac{1}{25}$.

Also $\frac{1}{25} = 0.04$ So our complete answer becomes 56.04.

Great! Hope that was easy?

Let us take the example of 7281 ÷ 129 and understand this better.

Step 1

So as usual, we write down the powers of 2, all the way up to 32, in the left-hand side column.

7281 ÷ 129

1
2
4
8
16
32

Step 2

We then start doubling our divisor which is 129 in the right-hand side column. So, we get 129, double of that is 258, then double of 258 is 516, double of 516 is 1032 and so on all the way up to 4128.

We stop at 4128 because if we double 4128, we will get 8256. This will be more than 7281 and we don't want that. So, our right-hand side column looks like this:

7281 ÷ 129

1	129
2	258
4	516
8	1032
16	2064
32	4128

Step 3

Now the objective on the right-hand side column is to find a series of numbers which add up to 7281. So, we get there by adding 4128 + 2064 + 1032 = 7224. But that's not 7281. So, like in the earlier example, the Egyptians had a way to deal with this.

We then subtract. And we get, 7281 – 7224 = 57, which is our remainder!

7281 ÷ 129

1	129
2	258
4	516
8 ←	1032
16 ←	2064
32 ←	4128

Also, note we have to consider the corresponding terms opposite of 4128, 2064 and 1032 in the left-hand column.

So, we get, 32 + 16 + 8 = 56, which is our quotient.

So, our final answer then becomes $56\frac{57}{129}$.

ACTIVITY 1: 2-DIGIT DIVISORS

Find the quotient.

1. 5,994 ÷ 54 = **2.** 5,915 ÷ 91 = **3.** 6,700 ÷ 67 =

4. 9,471 ÷ 33 = **5.** 5,264 ÷ 56 = **6.** 6,734 ÷ 74 =

7. 5,684 ÷ 98 = **8.** 7,200 ÷ 40 = **9.** 5,047 ÷ 49 =

10. 8,080 ÷ 80 = **11.** 4,080 ÷ 80 = **12.** 2,989 ÷ 61 =

13. 4,864 ÷ 64 = **14.** 4,165 ÷ 35 = **15.** 6,612 ÷ 76 =

16. 9,920 ÷ 62 = **17.** 5,742 ÷ 99 = **18.** 4,416 ÷ 92 =

19. 9,361 ÷ 37 = **20.** 6,384 ÷ 76 = **21.** 3,471 ÷ 89 =

22. 6,380 ÷ 55 = **23.** 7,866 ÷ 46 = **24.** 5,310 ÷ 45 =

25. 7,047 ÷ 87 =

ACTIVITY 2: 3-DIGIT DIVISORS

Find the quotient.

1. 9,452 ÷ 556 = **2.** 5,957 ÷ 851 = **3.** 2,955 ÷ 985 =

4. 8,606 ÷ 662 = **5.** 9,548 ÷ 682 = **6.** 3,040 ÷ 190 =

7. 9,344 ÷ 584 = **8.** 7,144 ÷ 376 = **9.** 3,560 ÷ 890 =

10. 9,061 ÷ 697 = **11.** 5,394 ÷ 899 = **12.** 5,688 ÷ 948 =

13. 7,518 ÷ 537 = **14.** 9,581 ÷ 871 = **15.** 4,980 ÷ 830 =

16. 3,278 ÷ 298 = **17.** 6,688 ÷ 418 = **18.** 4,680 ÷ 234 =

19. 3,848 ÷ 962 = **20.** 7,824 ÷ 652 = **21.** 6,540 ÷ 109 =

22. 3,900 ÷ 150 = **23.** 5,466 ÷ 911 = **24.** 5,840 ÷ 365 =

25. 3,808 ÷ 544 =

If I were to ask you where you would find the following—Merlion, Supertrees, Marina Bay Sands Skypark, Sentosa and the world's best maths curriculum—you would be right if, on the map, you place your finger on Singapore! Singapore, since its independence from the British, has become one of the world's most prosperous countries.

It has the world's best and most elite maths curriculum as per the Trends in International Mathematics and Science Study (TIMSS, https://nces.ed.gov/timss/). This is a study done globally over the years ranking countries based on their maths and science scores. Singapore has consistently ranked at the top, and countries like the United States and the United Kingdom have ranked in the twenties out of over seventy-five countries.

So, what makes the Singapore maths method so special that it is always numero uno? Singapore maths focuses on quality learning, rather than quantity. They will learn less but learn in great detail, thus having a strong foundation. So, let us see an example in division—the way they would crack it in Singapore!

Say, we have to do this division, $442 \div 2$.

We can break this up by place values of hundreds, tens and units.

So, we have $442 = 400 + 40 + 2$.

Now we divide each figure by 2.

So, we will have $\frac{400}{2} = 200$, then we will have $\frac{40}{2} = 20$ and finally $\frac{2}{2} = 1$.

Please note, here we have carefully chosen 400 as it is easy to divide 400 by 2 to give us 200. We are breaking up 442 as much as possible so that it is divisible by 2. Here we break it up as 400 because 400 is divisible by 2. This can be done in other ways also.

We then add the quotients together to get $200 + 20 + 1 = 221$. This is our final answer.

Let's take another problem $750 \div 3$

This time because we have the divisor as 3, we break the number 750 in an interesting way. So, we have $750 = 300 + 300 + 150$. Please note, here we have carefully chosen 300 as it is easy to divide 300 by 3 to give us 100. We are breaking up 750 as much as possible so that each part is divisible by 3. Here we break it up in terms of 300 first because 300 is divisible by 3. This can be done in other ways also.

So, we will have $\frac{300}{3} = 100$, then we have again $\frac{300}{3} = 100$ and finally we will have $\frac{150}{3} = 50$. We then add all the quotients together to get $100 + 100 + 50 = 250$, our final answer.

Now that was simple enough and very straightforward, right?

When I was looking at Singapore maths, I noticed that its hallmark is a very visual and pictorial approach to solve word problems. Most of my students were very scared of these kinds of problem as they would not understand how exactly to solve them. Therefore, when I read about the Singapore maths visual approach to solving word problems, I was very intrigued. It is a novel way called 'Singapore maths bar model strategy' and I will now share it with all of you.

The 'Singapore maths bar model strategy' is of two types. The part–whole model and the comparison model. Attempt this with an open mind, as this method is different from the traditional methods of working out word problems, and you will see in time, much easier too!

Let's take a word problem and understand the part–whole model first.

Miraaya purchased a monthly cable TV subscription for 5 months for $50. What was the monthly subscription?

Step 1

Here we draw the bar which represents the whole or $50.

Step 2

We break it up into five parts representing the purchase done for 5 months.

Step 3

Here we know the whole which is \$50 and the number of parts (months) which is 5. So, to find one part or monthly subscription, we divide $50 \div 5 = \$10$. Hope this is clear by the visual bar model? The bar model has been given so that you understand the word problem visually and the actual calculation of the division will be done the normal way.

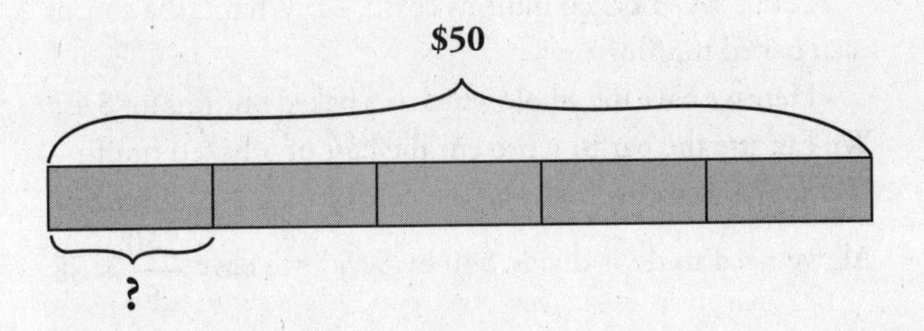

Let's take another word problem.

Bryan reads 12 books every month. By when would he be able to read 72 books?

Here we can see after reading the problem, that the whole is 72 books. So, we draw the bar representing the whole or 72 books.

Now each month Bryan reads one part of the whole or 12 books. So to figure out how many months it will take him to read 72 books, we divide $\frac{72}{12} = 6$. So, our answer is 6 months.

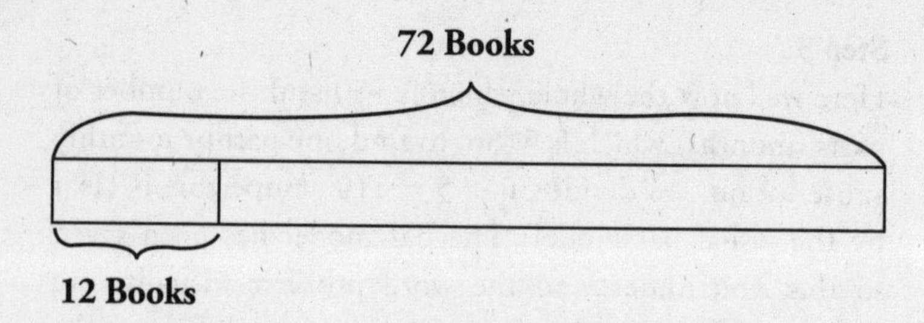

Let us take another word problem.

Let us say, 5 baked muffins cost $ 40. What is the cost of each baked muffin?

Here we have the whole which is 5 baked muffins for $ 40. We can use the bar to represent the cost of 5 baked muffins. We simply have to find out the cost of each baked muffin. All we need to do is divide $40 by 5. So, we have $\frac{\$40}{5} = \$8$

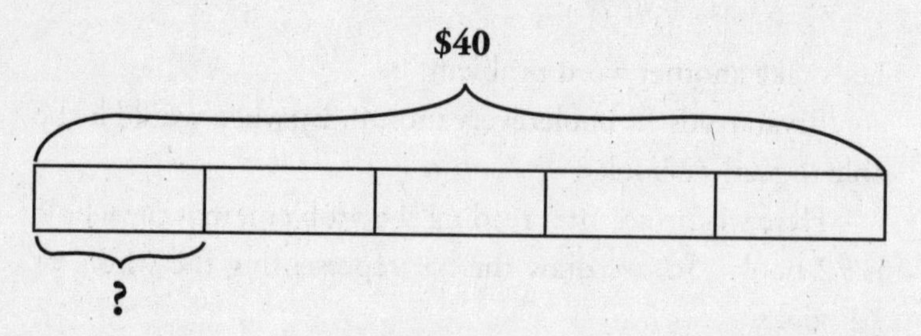

We can now look at the comparison model. Let's take a word problem.

In a seaport, there are 27 boats. And there are 3 times as many boats as yachts. How many yachts are there?

Here as you can see there are two quantities which are being compared—boats and yachts.

So, first we draw the bar representing 27 boats.

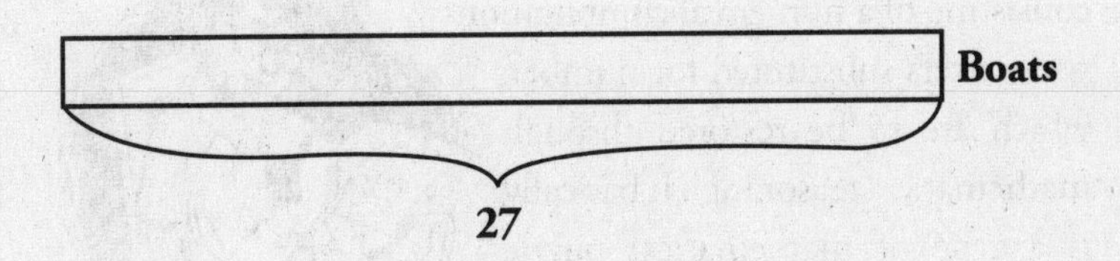

Thereafter, we see that there are 3 times as many boats as yachts. So, we can draw the bar model like this:

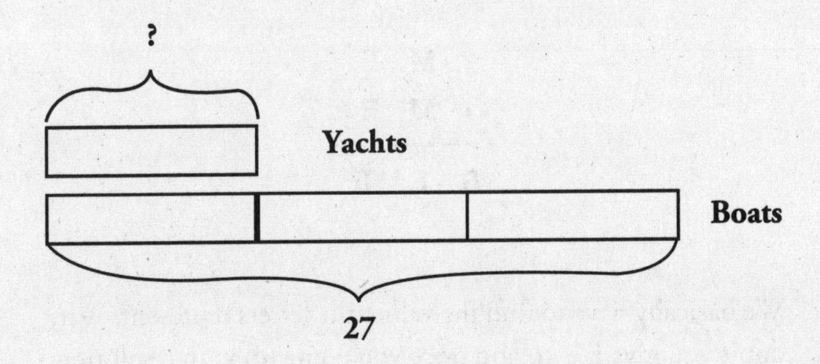

Now we simply divide 27 with 3 to find out the number of yachts.

The number of yachts becomes $\frac{27}{3} = 9$

Alphabets + Arithmetic = Alphametics

The Merriam Webster dictionary describes alphametics as 'a mathematical puzzle consisting of a numerical computation with letters substituted for numbers which are to be restored through mathematical reasoning'. It basically is a type of mathematical puzzle in which the digits in numerical calculations are replaced by letters of the alphabet.

Let's look at a very simple alphametic:

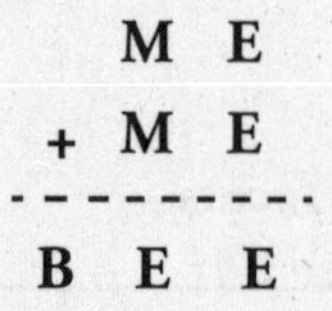

We basically have to find the values the letters represent. Why don't you give it a go and once you come up with a solution, you can come back here and check the solution.

Step 1

Note that in this alphametic, the column of the unit's digits is: E + E = E. There is only one digit, which has the property that when you add it to itself, you get the same digit as the result and that is zero! Only the sum of two zeros is zero, so E must be equal to 0. So, our alphametic looks like this so far:

```
   M 0
 + M 0
---------
 B 0 0
```

Step 2

The letter B must represent the digit 1, since when you add two 2-digit numbers you cannot possibly get a number larger than 198. Even you get 198 when both addends are 99. Since M and E are two different numbers, they will certainly be even smaller than 99! In any case, the hundreds digit in the sum, represented by B in our example, must be 1. Our alphametic looks like this:

```
   M 0
 + M 0
---------
 1 0 0
```

Step 3

Now can you tell me a number which when added to itself gives us 100? So, M has to be 5. We will have now in our alphametic 50 + 50 = 100. This is our solution. Wasn't it rather easy? Our final and complete alphametic looks like this:

```
   5 0
 + 5 0
---------
 1 0 0
```

I hope you enjoyed it! Let me now share another example with you. This one is extremely simple and with your new mathematical reasoning skills, you should be able to crack this one easily! So here we go:

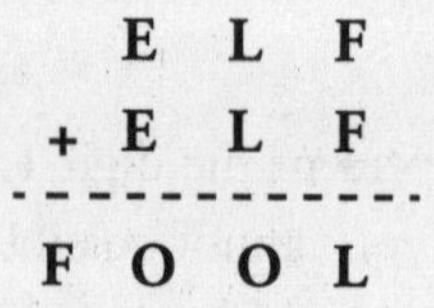

Step 1

This is a three digit plus a three-digit number, giving us a four-figure answer. Even if we take the two numbers to be the highest three-digit figure which is 999, the leftmost digit in our alphametic would be 1.

The idea here is, if we add the two highest three digit numbers which is 999 + 999 it will give us 1998. So see the leftmost digit will always be 1.

So, in this case F = 1 and that is the starting point of our alphametic, which now looks like this:

```
   E L 1
 + E L 1
 -------
 1 O O L
```

Step 2

If we now look at the rightmost column, we have 1 + 1 = 2. So, L = 2. So now our alphametic in the middle is 2 + 2 = 4. So, O = 4. Our alphametic now looks like this:

```
    E  2  1
 +  E  2  1
------------
 1  4  4  2
```

Step 3

In our final step, we will find out the value of E. In perspective, since all the values of the other letters are known, we see that E + E = 14. So, we have E = 7.

Our solved alphametic looks like this:

```
    7  2  1
 +  7  2  1
------------
 1  4  4  2
```

Fascinating stuff, isn't it? I came across alphametics when I was preparing for my business school examination more than eighteen years ago. I was inclined to do various puzzles and maths reasoning problems and that's where my journey into mathematics began—but that's a story I will save for another day! I will now share three more very popular alphametics for you to enjoy!

```
1.                2.                 3.
   O O O H            E A T            S E N D
 + F O O D        + T H A T          + M O R E
----------        ----------         ----------
 F I G H T        A P P L E          M O N E Y
```

5

The 'Digit Sum' to Check Your Answers

Digit Sums

The concept of 'digital sums' is very useful as it helps us check whether our answers are correct or not. The word 'digit' refers to numbers like 1, 2, 3, 4, 5 etc. And the word 'sum' means 'to add'. So, combining the two we have 'digit sum' which is nothing but the sum of the digits. Here we simply have to add the digits in the number till the time we have reached a single figure. Let's take an example!

Say we have to find the digit sum of 43, it is nothing but $4 + 3 = 7$, our answer is 7.

Say now we have to find the digit sum of 92. We add $9 + 2 = 11$. We add again $1 + 1 = 2$. So, 2 is the digit sum of 92. Hope this is clear.

Now let's take a look at the digit sums of few more numbers.

Number	Summing Digits	Digit Sum
48	4 + 8 = 12 = 1 + 2 = 3	3
645	6 + 4 + 5 = 15 = 1 + 5 = 6	6
8541	8 + 5 + 4 + 1 = 18 = 1 + 8 = 9	9
25316	2 + 5 + 3 + 1 + 6 = 17 = 1 + 7 = 8	8
447501	4 + 4 + 7 + 5 + 0 + 1 = 21 = 2 + 1 = 3	3
10345	1 + 0 + 3 + 4 + 5 = 13 = 1 + 3 = 4	4
990214	9 + 9 + 0 + 2 +1 + 4 = 25 = 2 + 5 = 7	7

Now let us see another method of finding the digit sum of a number. The method is called 'casting out nines' and it helps us to compute the digit sum in a quick manner. Once we've understood this, we shall see the process to check the answers. So, let's get started.

Casting Out Nines

Under this method, we find out the digit sum by simply casting out 9 and those digits which add up to 9. For example, say if we have to find out the digit sum of 652109, we will simply cast out 9 and 6, 2 and 1 as they add up to 9. Our sum would look like this ~~6~~5~~21~~0~~9~~. So now we have 50 which remains and which adds up to 5 + 0 = 5. So, our digit sum of 652109 is 5.

Let's take another example 5732124. Here, we cancel those digits which add up to 9. So, we cancel out 5 and 4 and 7 and 2. Our sum now looks like ~~57~~321~~24~~. We are left with 321 = 3 + 2 + 1 = 6. So, our digit sum for 5732124 is 6.

The 9-Point Circle

I would now like to share with you the concept of the 9-point circle. The concept of a 9 point circle is very important as it has many uses.

Let's write down the digit sums of natural numbers from 1 onwards.

No	1	2	3	4	5	6	7	8	9	**10**	11	12	13	14	15	16	17	18	19
Digit Sums	1	2	3	4	5	6	7	8	9	**1**	**2**	**3**	**4**	**5**	**6**	**7**	**8**	**9**	**1**

Note the sequence. It is from 1 to 9. We can now represent this sequence of the digit sums in a 9-point circle, as shown here.

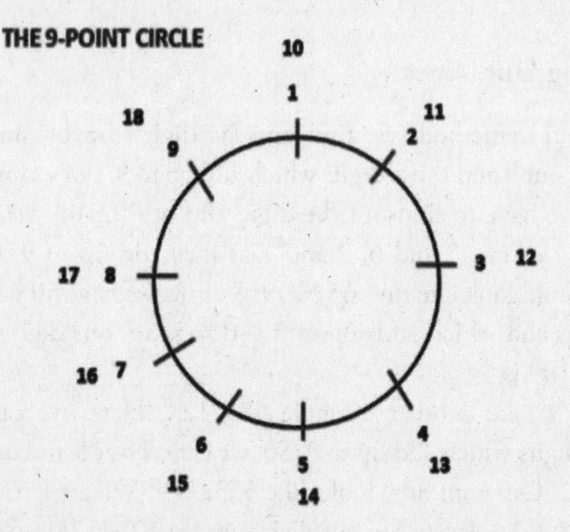

Let's take any branch, say the first branch—1—for example. All the numbers in this branch will have the digit sum to be 1. So, the digit sum of 10 is 1+0=1. Let's take another branch, say 7. All the numbers in this branch too will have the digit sum to be 7.

There is also another thing to learn from the 9-point circle. We learn that if we add 9 to any number or subtract 9 from any number, it does not affect its digit sums.

Also in the digit sum, 9 and 0 are considered to be equivalent. Let's take the number 8136 and find its digit sum. We have 8 + 1 + 3 + 6 = 18 = 1 + 8 = 9. Also, if we cast out nines, we get 0. So, to say that the digit sum of a number is 9 or 0, essentially means the same thing!

Using Digit Sums to Check Answers

Now that we have understood the concept of digit sums, we can move ahead and see how we can check our maths problems with this system.

Checking Additions

Let's begin with addition, taking an example, 233 + 178

The sum of these two numbers is 411; let us now check and see whether this is the correct answer or not.

Step 1

We first find out the digit sum of 233. It is 2 + 3 + 3 = 8. We then find the digit sum of 178, which is 1 + 7 + 8 = 16

= 1 + 6 = 7. Adding these two-digit sums, we get 8 + 7 = 15 = 1 + 5 = 6.

		Digit Sum
233	⟶	8
+ 178	⟶	+ 7
411	⟶	6

Step 2

We then find the digit sum of our answer 411 which is 4 + 1 + 1 = 6. Now since both the digit sums match, we can say that our answer 411 is correct.

Let's take another example to understand this better. Say we have 57461 + 90146.

Step 1

		Digit Sum
57461	⟶	5
+ 90146	⟶	+ 2
147607	⟶	7

We first find out the digit sum of 57461, which is 5 + 7 + 4 + 6 + 1 = 23 = 2 + 3 = 5. We then find out the digit sum of 90146 = 9 + 0 + 1 + 4 + 6 = 20 = 2 + 0 = 2. Adding the two-digit sums, we get 5 + 2 = 7.

Step 2

Finally, we take the digit sum of 147607. It is $1 + 4 + 7 + 6 + 0 + 7 = 25 = 2 + 5 = 7$. Now, since both the digit sums match, we can say that our answer 147607 is correct.

Checking Subtractions

We'll now check subtractions using digit sums. Let's take an example say 65141 – 25896. Let's check it now!

		Digit Sum
65141	⟶	8
– 25896	⟶	– 3
39245	⟶	5

Step 1

The digit sum of 65141 is 8 and the digit sum of 25896 is 3. The difference in these two digit sums is $8 - 3 = 5$.

Step 2

The answer we get is 39245, which has a digit sum of 5. So, we note that both the digit sums are the same and hence conclude that our answer is correct!

Let us take another example 4321 – 1786.

		Digit Sum
4321	⟶	1
– 1786	⟶	– 4
2535	⟶	– 3

Step 1

We find the digit sum of 4321, which is 1. We then find the digit sum of 1786, which is 4. We subtract the digit sums 1 – 4 and we get –3. Don't worry we can also get digit sums in a negative form and we can then make it positive by adding 9. So, this becomes 9 – 3 = 6.

Remember, adding or subtracting 9 from any number does not change the digit sum. Because this is a negative digit sum, we add 9 and get 6 as our final digit sum.

Step 2

The digit sum of our answer 2535 is 6, which matches with the digit sum above. So, we can say with confidence that our answer 2535 is correct.

Checking Multiplications

Let's check the example 56 × 84.

Step 1

We first calculate the digit sum of 56, which is 2 and then we compute the digit sum for 84, which is 3. We then multiply these two digit sums which gives us 6.

		Digit Sum
56	⟶	2
× 84	⟶	× 3
4704	⟶	6

Step 2

We finally calculate the digit sum of our answer 4704 which is also 6. So, we can safely conclude that our answer is absolutely correct.

Now let's check the example 258 × 471.

Step 1

We find the digit sum of 258 which is 6 and that of 471, which is 3. We then multiply these digit sums which gives us 6 × 3 = 18. Now the digit sum of 18 is 9.

		Digit Sum
258	⟶	6
× 471	⟶	× 3
121,518	⟶	9

Step 2

We compute our answer to be 121,518. We then calculate the digit sum of 121518 to be also 9. Making our computation absolutely correct!

A Word of Caution

The digit sum is an important checking tool, but it has its limitations as well.

For example, let's take 12 × 34.

Supposing we write our answer as 804 instead of the correct answer 408. The digit sum of 804 is 8 + 0 + 4 = 12 = 1 + 2 = 3. But when we check our answer, it gives us 408. Both 804 and 408 have the same digit sum of 3. This is a limitation of the digit sum method. Please take care to write the answer in the correct order or else, an error may occur.

ACTIVITY: MATHEMATICAL PATTERNS IN A VEDIC SQUARE

Let us explore a Vedic square. A Vedic square is a multiplication table that gives us various mathematical patterns.

Let's complete this multiplication table:

1	2	3	4	5	6	7	8
2							
3							
4							
5							
6							
7							
8							

Step 1

Just write down the multiplication tables of 2 in the second row, 3 in the third row, 4 in the fourth row and so on. I have completed a part of the table, and you complete the rest.

1	2	3	4	5	6	7	8
2	4	6	8	10	12	14	16
3	6	9	12	15	18	21	24
4	8	12	16	20	24	28	32
5							
6							
7							
8							

Step 2

So, in a Vedic square, you just write down the digit sum of the numbers in the multiplication table. You get the digit sum of two-digit numbers by adding the digits together until you get a single digit answer:

- So, for the number 14, the digit sum is 1 + 4 = 5
- For the number 28, the digit sum is 2 + 8 = 10, then 1 + 0 = 1. Complete this Vedic square using the multiplication table above:

1	2	3	4	5	6	7	8
2							
3							
4							
5							
6							
7							
8							

After completion, our Vedic square will look like this:

1	2	3	4	5	6	7	8
2	4	6	8	1	3	5	7
3	6	9	3	6	9	3	6
4	8	3	7	2	6	1	5
5	1	6	2	7	3	8	4
6	3	9	6	3	9	6	3
7	5	3	1	8	6	4	2
8	7	6	5	4	3	2	1

Mathematical Patterns in the Vedic Square

Here are six Vedic squares. Find in each Vedic square, the given number every time it appears in the square, then shade them up and join them to create a shape. I've done the first one as an example.

Find: 1

1	2	3	4	5	6	7	8
2	4	6	8	**1**	3	5	7
3	6	9	3	6	9	3	6
4	8	3	7	2	6	**1**	5
5	**1**	6	2	7	3	8	4
6	3	9	6	3	9	6	3
7	5	3	**1**	8	6	4	2
8	7	6	5	4	3	2	**1**

Find: 2

1	2	3	4	5	6	7	8	9
2								
3								
4								
5								
6								
7								
8								
9								

Find: 3

1	2	3	4	5	6	7	8	9
2								
3								
4								
5								
6								
7								
8								
9								

Find: 4

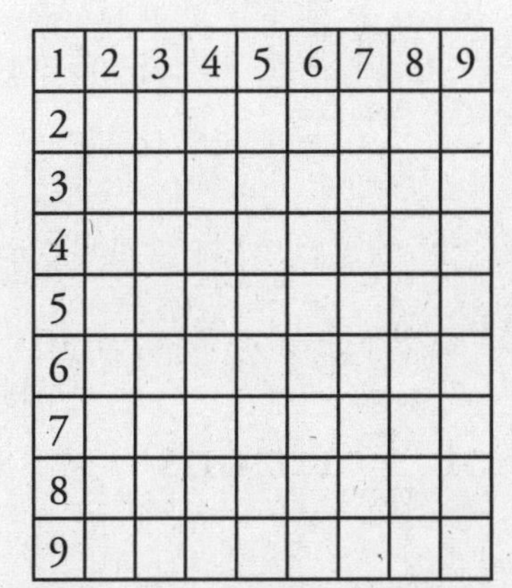

1	2	3	4	5	6	7	8	9
2								
3								
4								
5								
6								
7								
8								
9								

Find: 5

1	2	3	4	5	6	7	8	9
2								
3								
4								
5								
6								
7								
8								
9								

Find: 6

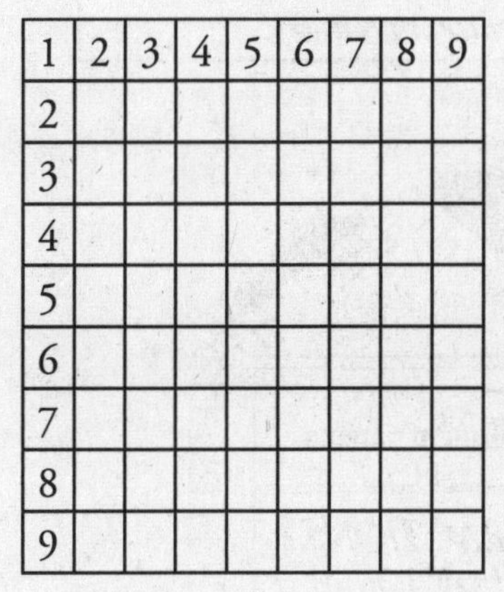

1	2	3	4	5	6	7	8	9
2								
3								
4								
5								
6								
7								
8								
9								

Find: 7

1	2	3	4	5	6	7	8	9
2								
3								
4								
5								
6								
7								
8								
9								

6

No More Fear of 'Fractions'

Did you know that fractions as we use them today didn't exist in Europe until the seventeenth century?

Students all over the world get really worked up when it comes to fractions. Or if I may say it, they freak out when it comes to understanding and solving word problems related to fractions. The Singapore maths way to do fraction word problems is very quick, simple and easy. Here I have applied the 'part–whole bar model' method, which we learnt in the chapter on division. So, let's go ahead and work out some word problems on fractions.

1) Vaani and Leela bought two similar cakes. Vaani ate $\frac{1}{2}$ of her cake and Leela ate $\frac{1}{4}$ of her cake. Who ate a bigger portion of their cakes?

Let us represent the fractions using bars. The shaded parts show the portion of the cake they have eaten.

Vaani ate $\frac{1}{2}$ of her cake:

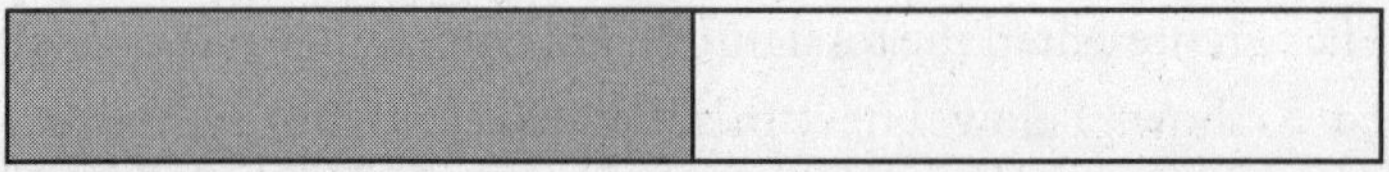

Leela ate $\frac{1}{4}$ of her cake:

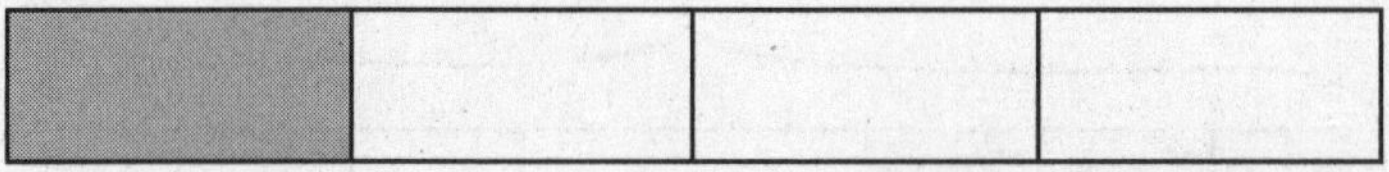

We can clearly see that Vaani ate more cake than Leela.

2) If the cost of one-fourth of a kilogram of mangoes is $9, what is the cost of a kilogram of mangoes?

Here, we represent the kilogram of mangoes as a bar with four parts. One unit or one fourth is \$9.

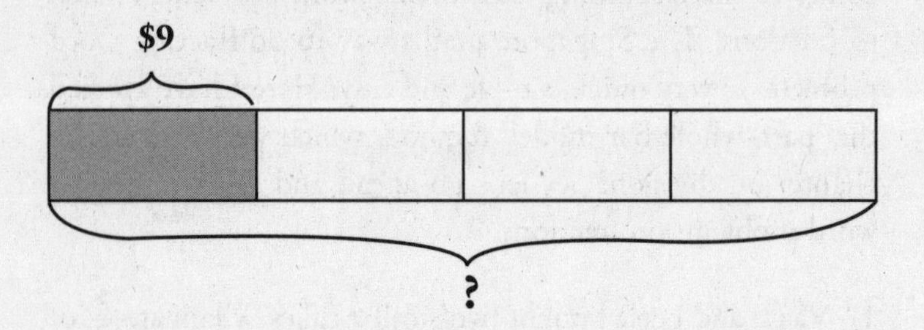

Therefore 4 units would be $\$9 \times 4 = \$36$. So, our answer is \$36.

3) $\frac{3}{8}$ of a group of animals in a circus were lions. If there were 30 lions, how many animals were there in all?

I have represented the total number of animals in a rectangular bar as shown below. I have then divided them into eight equal parts.

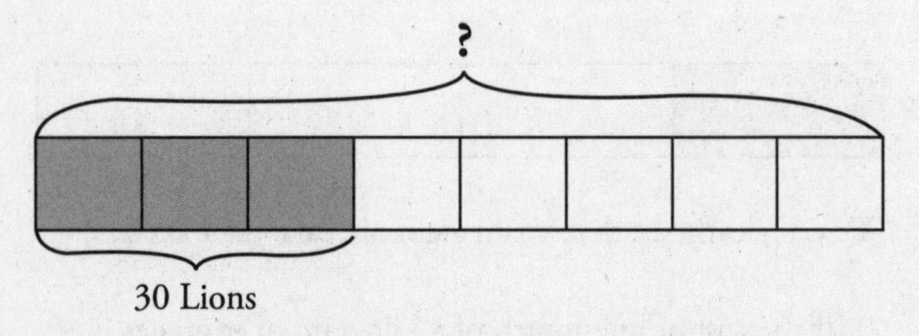

Now since $\frac{3}{8}$ are lions, we shade 3 out of the 8 parts. Now the 3 parts (out of the 8) represent the total number of lions, which is 30. So, one part would be $\frac{30}{3} = 10$. And we need 8 parts, so we simply multiply $10 \times 8 = 80$ animals. Our answer is that there are 80 animals in all.

4) Manish was gifted some postcards by his grandfather. He gave $\frac{2}{7}$ of the postcards to Raj. If he gave 20 postcards to Raj, how many postcards did Manish have at first?

Let us represent the total number of postcards as a rectangular bar. We also divide the bar into seven parts. Our bar looks like this:

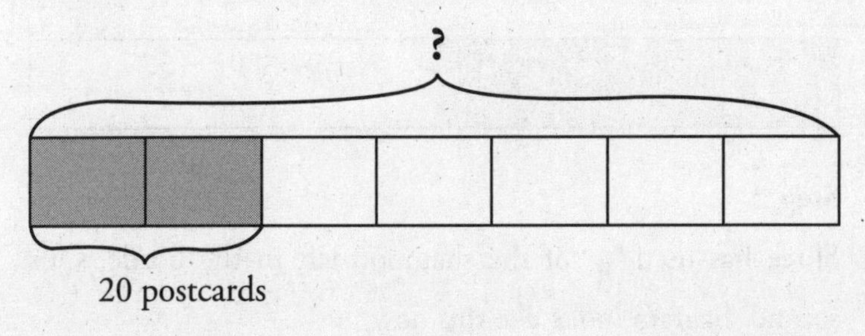

Here, we can clearly see that 20 postcards that have been given to Raj form $\frac{2}{7}$ of the total number of postcards. So, we divide $\frac{20}{2}$ to give us 10 for each part, as per the diagram. So, for seven parts or the total number of postcards, we have $10 \times 7 = 70$ postcards. So, 70 postcards are our final answer.

5) Shree has a half-filled bottle of shampoo. She uses $\frac{1}{3}$ of the shampoo left in the bottle.
   a) How much shampoo is still there in the bottle?
   b) If she now buys a shampoo sachet and fills one sixth of the bottle, how much shampoo remains in the bottle now?

Step 1

Let us represent the shampoo in the bottle as a bar model. We shade $\frac{1}{2}$ of the bar because $\frac{1}{2}$ of the bottle is filled with shampoo. Our diagram looks like this:

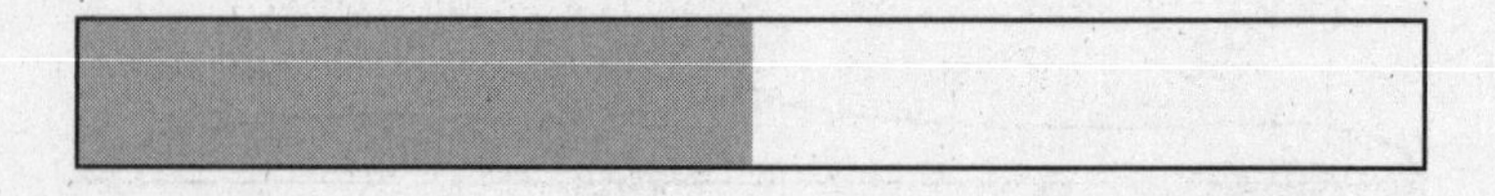

Step 2

Shree has used $\frac{1}{3}$ of the shampoo left in the bottle. Our second diagram looks like this now:

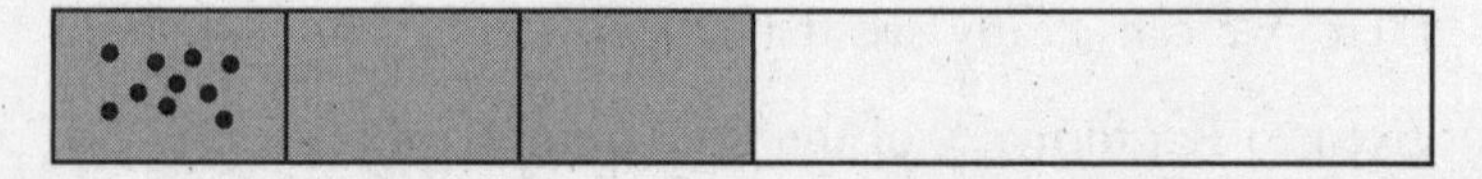

Here above you can see that Shree has used $\frac{1}{3}$ of the shampoo left represented by the dots.

Step 3

So, when we compare the two, we need to answer, how much shampoo is left in the bottle.

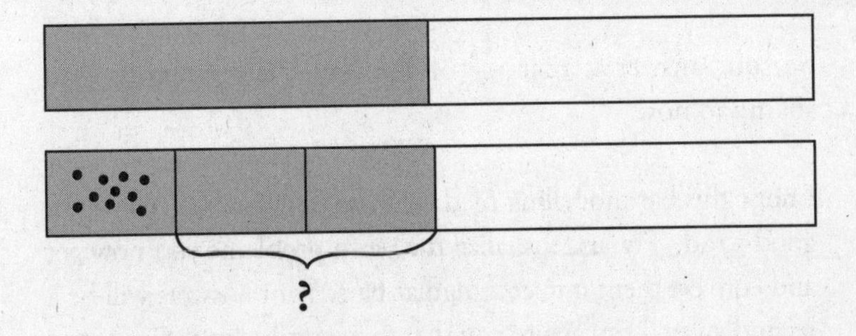

So now we would basically calculate

$$\frac{1}{2} - \frac{1}{6} = \frac{6-2}{12} = \frac{4}{12} = \frac{1}{3}$$

So that's our first answer.

Step 4

We can now draw a new diagram with the remaining shampoo $\frac{1}{3}$, like this:

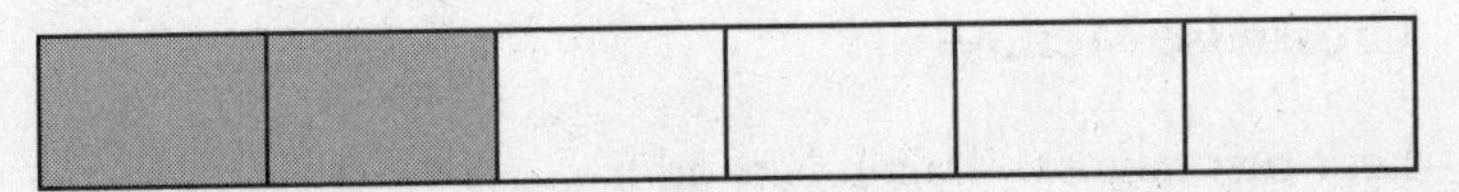

Step 5

We add another $\frac{1}{6}$ of the bottle with shampoo, so it becomes

$$\frac{1}{3} + \frac{1}{6} = \frac{6+3}{18} = \frac{9}{18} = \frac{1}{2}$$

Our diagram looks like this:

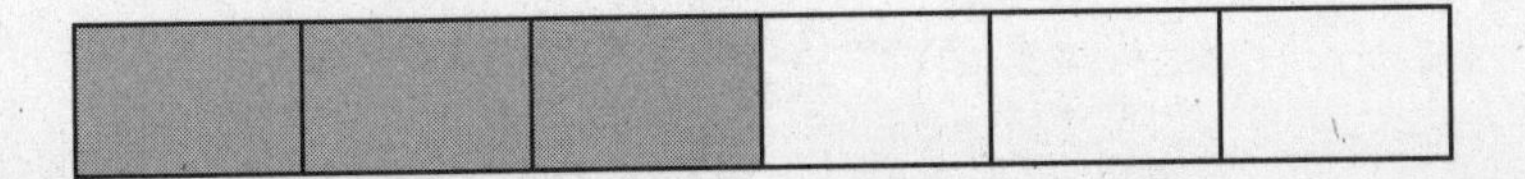

So, our answer is that $\frac{1}{2}$ of the bottle is remaining with shampoo now.

I hope this bar modelling method from Singapore keeps you in good stead. Try and visualize the word problems you now get and convert them into rectangular bars. Your answers will be a second away, I can assure you that as a promise from Singapore!

Addition and Subtraction of Fractions

Do you remember the 'vertically and crosswise' method for multiplication of two-digit numbers? Well, this method is important as it is helpful in a lot of topics. One of them is fractions. We can use the method, vertically and crosswise, for adding and subtracting fractions as well. Let's see how.

Addition of Fractions Using Vertically and Crosswise Method

Let's first take two fractions to add:

$$\frac{3}{5} + \frac{1}{4}$$

Step 1

Here, we'll first multiply crosswise and then add the product, as seen in the diagram below. So, that gives us $(4 \times 3) + (5 \times 1) = 17$. This becomes our numerator

Step 2

Now, we will work out our denominator simply by multiplying 5 and 4. So, $5 \times 4 = 20$. So, our complete answer now becomes $\frac{17}{20}$.

Our sum would look like this:

$$\frac{3}{5}+\frac{1}{4}$$

$$=\frac{12+5}{20}$$

$$=\frac{17}{20}$$

Now, let's solve another quick example: $\frac{2}{11}+\frac{7}{9}$

Step 1

In the first step, we'll multiply crosswise as shown here:

$$\frac{2}{11} \times \frac{7}{9}$$

So that equals $(9 \times 2) + (11 \times 7) = 18 + 77 = 95$.

Now this 95 becomes our numerator. Let's find the denominator in the second step.

Step 2

In the second step, we'll get our denominator by multiplying 11 and 9. So, 11 × 9 = 99. Now, this 99 becomes our denominator.

$$\frac{2}{11} \times \frac{7}{9}$$

And our sum now looks like this

$$= \frac{(2 \times 9) + (11 \times 7)}{11 \times 9} = \frac{95}{99}$$

Let's look at another example: $\frac{3}{8} + \frac{1}{12}$

Step 1

As done in the earlier examples, here too, we'll first multiply crosswise as shown here.

$$\frac{3}{8} \times \frac{1}{12}$$

So now that gives us (12 × 3) + (8 × 1) = 36 + 8 = 44. This becomes our numerator.

Step 2

Step 2 will give us our denominator. So, we have: 12 × 8 = 96

$$\frac{3}{8} \times \frac{1}{12}$$

So our sum now looks like this:

$$\frac{3}{8}+\frac{1}{12}=\frac{36+8}{96}=\frac{44}{96}=\frac{11}{24}$$

We got $\frac{44}{96}$, which we simplified to $\frac{11}{24}$ and that is our answer.

So this was for addition, now let's look at subtractions of fractions.

Subtraction of Fractions Using the Vertically and Crosswise Method

Just as we saw earlier with addition, the vertically and crosswise method plays an important role in subtraction as well. We can subtract any fraction from any other fraction using this formula, and very easily too!

Let's take an example $\frac{6}{7}-\frac{1}{2}$

Step 1

We'll first find the numerator. All we got to do is multiply crosswise.

So we have $(6 \times 2) - (7 \times 1) = 12 - 7 = 5$, so 5 is our numerator.

$$\frac{6}{7} \times \frac{1}{2}$$

Step 2

Now we'll find out the denominator. Which is just multiplication of the denominators. So, $7 \times 2 = 14$.

$$\frac{6}{7} \times \frac{1}{2}$$

Our sum looks like this:

$$\frac{6}{7}-\frac{1}{2}=\frac{12-7}{14}=\frac{5}{14}$$

So our answer becomes $\frac{5}{14}$.

Let's take another example $\frac{12}{25}-\frac{3}{50}$

Step 1

Here too, as before, we'll be applying the vertically and crosswise method to the problem.

Firstly, we multiply crosswise to get our numerator, as shown here:

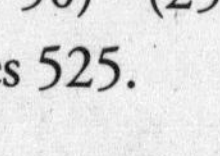

$$\frac{12}{25} \times \frac{3}{50}$$

Now we'll subtract: (12 × 50) –(25 × 3) = 600 – 75 = 525. So our numerator becomes 525.

Step 2

Next, we'll find out the denominator. So that will be: 25 × 50 = 1250.

$$\frac{12}{25} \times \frac{3}{50}$$

Our sum looks like this:

$$\frac{12}{25}-\frac{3}{50}=\frac{600-75}{25\times50}=\frac{525}{1250}=\frac{21}{50}$$

This is our answer.

Let's work on another example: $\frac{6}{11}-\frac{3}{7}$

Step 1

Our first step will be to multiply crosswise.

$$\frac{6}{11}\times\frac{3}{7}$$

We will subtract: (6 × 7) – (11 × 3) = 42 – 33 = 9

So 9 becomes our numerator!

Step 2

Here, we'll work to get out denominator.

We'll now multiply the two denominators to get (7 × 11) = 77

$$\underbrace{\frac{6}{11}\times\frac{3}{7}}$$

$$\frac{6}{11}-\frac{3}{7}=\frac{42-33}{77}=\frac{9}{77}$$

So our answer becomes $\frac{9}{77}$ as shown here! Just a finger snap and there comes the answer in a maximum of two steps.

Did you know?

The Egyptians were using fractions from as early as 1800 BC. But they wrote all their fractions using unit fractions. A unit fraction has 1 as its numerator. All the other fractions were the sum of unit fractions, without any repetition.

So, we would have:

$\frac{1}{2} = \frac{1}{3} + \frac{1}{6}$ Or $\frac{1}{5} = \frac{1}{6} + \frac{1}{30}$

So, having said that, can you now express $\frac{1}{3}$ as the sum of two unit fractions?

7

The Magic of Magic Squares!

In this chapter, we will learn various methods to solve *squares*. A square is a number multiplied by itself. Say, if we have to find the square of 7, we will express it as 7^2. This means that 7 is multiplied by 7 to give us 49. I hope this concept of squares is clear to you.

A) Squares of Numbers Ending in 5

I will now share a method with you to quickly and mentally find the squares of numbers ending in 5. Here, we will apply a special maths sutra from India called 'by one more than one before'.

Say I ask you to square the number 5, it is 25 and now I ask you to square 15, which equals 225. And again, if I ask you for 25^2, the answer will be 625. Let me list it down for you like this:

$5^2 = 25$
$15^2 = 225$
$25^2 = 625$

Now friends, do you see a pattern here? If not, let me share it with you.

See, whenever you square any number ending in 5, the last two digits of our answer will always be 25. And to get to the digit(s) prior to this, we apply the formula 'by one more than one before'.

So, in case of 15^2, we will multiply 1 with one more than one before, which is 2. So $1 \times 2 = 2$.
So, our answer becomes 225.

Let's now take 35^2.
As usual, we can simply write the last two digits to be 25. For the digit(s) prior to this, we have 3 and one more than 3 is 4. So, we multiply 3×4 to give us 12. So, our answer becomes 1225.

Let's take one more, say 65^2.
Any guesses for the last two digits? Yes, 25! For the digit(s) prior to this, we multiply 6 with 7 (as 7 is one more than 6) to get 42. So, our final answer becomes 4225.

Now complete the table below:

Number	Square
5^2	25
15^2	225
25^2	
35^2	
45^2	
55^2	
65^2	
75^2	
85^2	
95^2	
105^2	
115^2	
125^2	

Note: For Squaring larger numbers ending in 5, like 225^2, we can multiply $22 \times 23 = 506$ and then add the 25 right at the end, giving us our answer 50625.

B) Squares of Numbers Close to a Base

We will now move on to solving squares of numbers close to a base like 10,100, 1000 and so on. We briefly touched upon it, in the chapter on multiplication—now let us explore it a bit more.

Say, we must find 101^2.

Step 1

Note that 101 is close to the base 100. The number 101 is more than 100 by + 01. So, we add this excess to 101, making the left-hand side part of our answer 102. Also, we take + 01 because 100 has two zeroes and there should be two digits on the right-hand side of our answer.

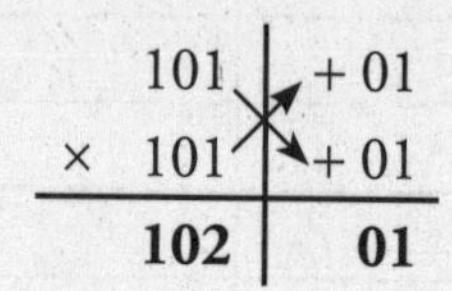

Step 2

Now since 100 is the base, there should be two digits on the right-hand side of our answer. So our excess is + 01. So, we square that and we get + 01 again. This leads to our answer 10201.

Was this clear? Let us now see another example.

Say we have to find out 102^2.

Let's do this mentally!

Step 1

So, 102 is close to our base 100. Also 102 is excess of 100 by + 02. We add + 02 to 102, to make it 104. This is our left-hand side part of the answer.

Step 2

To arrive at our final answer, we will now square + 02, which is + 04. Now + 04 is the right-hand side part of our answer. Now, combining the two, we get our final answer as **10404**.

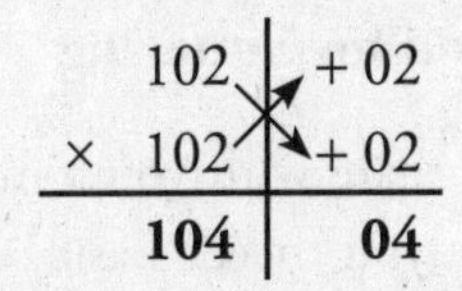

Now complete the table below:

Number	Square
101^2	10201
102^2	10404
103^2	
104^2	
105^2	
106^2	
107^2	
108^2	
109^2	
110^2	
111^2	
112^2	
113^2	
114^2	
115^2	

You can even extend this method to base 1000 or more. It always works!

C) General Method for Squaring

I would now like to share with you a more general method of squaring, where you need basic addition and basic multiplication skills. Anyone can use this technique to find squares of two-digit figures. It's simple and easy!

Let's take an example. Say we have to find 37^2.

Step 1

To start solving the sum 37^2, we square 3 first. We get 9. We put it down like this:

$$37^2 = \quad 9$$

Step 2

Next, we multiply 3 and 7. It gives us 21. We also write that down next to 9. It looks like this now:

$$37^2 = \quad 9 \quad 21$$

Step 3

We now square 7. It gives us 49. We place 49 next to 21, like this:

$$37^2 = \quad 9 \quad 21 \quad 49$$

Step 4

Now, all we have to do is put 21 in the middle, in the second line and add. We just copy our middle figure 21 in our first line below. Our problem now looks like this:

$$
\begin{array}{rrrr}
37^2 = & 9 & 21 & 49 \\
 & + & 21 & \\
\hline
\end{array}
$$

Step 5

We are now very close to getting our answer. So, once our sum is set up like this, our final step is to add. Remember each column will give us a single digit. So, we put down 9 from 49 to our rightmost answer digit place and carry 4 over to the next place to the left. We then add 21 + 21 + 4 (carry-over) = 46. We put 6 down and carry 4 to the left. Finally, we have 9 + 4 (carry-over) = 13. Our answer is 1369.

$$
\begin{array}{rrrr}
37^2 = & 9 & 21 & 49 \\
 & + & 21 & \\
\hline
 & 13 & 6 & 9
\end{array}
$$

Let's take a new example. Say we have to find 54^2.

Step 1

As before, to start solving the sum 54^2, we square 5 first. We get 25. We put it down like this:

$$54^2 = \quad 25$$

Step 2

Next, we multiply 5 and 4. It gives us 20. We also write that down next to 25. It looks like this now:

$$54^2 = \quad 25 \quad 20$$

Step 3

We now square 4. It gives us 16. We place 16 next to 20, like this:

$$54^2 = 25 \quad 20 \quad 16$$

Step 4

Now, all we have to do is put 20 from the top line in the middle, in the second line and add. Our problem now looks like this:

$$\begin{array}{rccc} 54^2 = & 25 & 20 & 16 \\ & + & 20 & \\ \hline \end{array}$$

Step 5

So, once our sum is set up like this, our final step is to add. We add from the right to the left, taking one digit at a time. From 16, we put down 6 as the rightmost answer digit and carry 1 over to the left. We then add 20 + 20 + 1 (carry-over) = 41. We put 1 down as the middle digit of our answer and carry 4 to the left. We finally have 25 + 4 (carry-over) = 29. Our final answer is 2916.

$$\begin{array}{rccc} 54^2 = & 25 & 20 & 16 \\ & + & 20 & \\ \hline & 29 & 1 & 6 \end{array}$$

ACTIVITY 1: EVALUATE

1. 70^2 = ___	**2.** 61^2 = ___	**3.** 57^2 = ___	**4.** 77^2 = ___	**5.** 20^2 = ___
6. 68^2 = ___	**7.** 64^2 = ___	**8.** 23^2 = ___	**9.** 24^2 = ___	**10.** 81^2 = ___
11. 45^2 = ___	**12.** 91^2 = ___	**13.** 32^2 = ___	**14.** 99^2 = ___	**15.** 38^2 = ___
16. 54^2 = ___	**17.** 19^2 = ___	**18.** 84^2 = ___	**19.** 41^2 = ___	**20.** 53^2 = ___
21. 71^2 = ___	**22.** 48^2 = ___	**23.** 92^2 = ___	**24.** 59^2 = ___	**25.** 73^2 = ___
26. 87^2 = ___	**27.** 88^2 = ___	**28.** 83^2 = ___	**29.** 89^2 = ___	**30.** 51^2 = ___
31. 75^2 = ___	**32.** 30^2 = ___	**33.** 46^2 = ___	**34.** 21^2 = ___	**35.** 78^2 = ___
36. 79^2 = ___	**37.** 58^2 = ___	**38.** 82^2 = ___	**39.** 62^2 = ___	**40.** 26^2 = ___
41. 22^2 = ___	**42.** 47^2 = ___	**43.** 98^2 = ___	**44.** 50^2 = ___	**45.** 31^2 = ___
46. 37^2 = ___	**47.** 80^2 = ___	**48.** 76^2 = ___	**49.** 40^2 = ___	**50.** 34^2 = ___

The Magic of Magic Squares

In a magic square grid, the numbers in each row, column and diagonal have the same constant sum—the constant magic sum. Let us see a magic square.

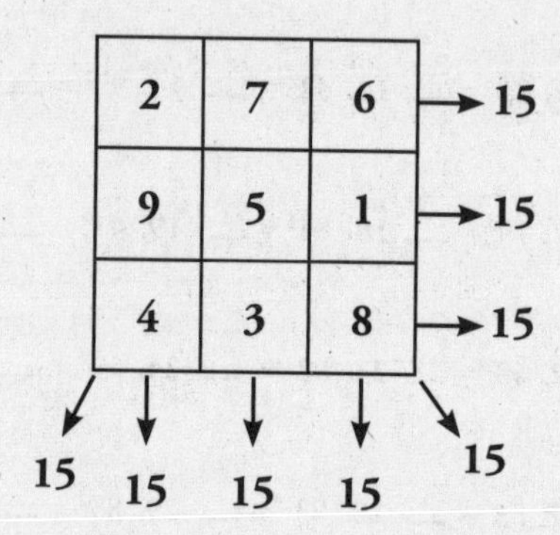

In the above magic square, we see that the rows, columns and diagonals have the same magic sum, which is 15. We also note that this is a magic square of order 3, as there are three rows and three columns. There can be magic squares of order 4 or more as well.

Magic squares have been around for over 4000 years or more. They have been found in the oldest of civilizations, including China, Egypt and India. People down the ages have believed that magic squares have mysterious properties. People believed that magic squares cured people of illness and hence helped them to live longer. Some people believed that magic squares attract wealth and luck and should be there in every house in India.

In India, the 3 × 3 magic square has been used as part of rituals for ages and if you visit the Parshvanath Jain temple in Khajuraho, you will see a popular 4 by 4 magic square from the tenth century. It is called Chautisa Yantra. Chautisa means 34 in Hindi and because the magic sum here is 34, it is called Chautisa Yantra!

ACTIVITY

Complete each magic square. Use any whole numbers. Each number can be used no more than once in each magic square.

1.

The sum is 15.

2		
	5	3

2.

The sum is 60.

32		
		28
		8

3.

The sum is 30.

	10	6
12		

4.

The sum is 60.

		16
28		
24		

8

Destination 'Percentages'

Did you know?

In ancient Rome, in the land of the great king Julius Caesar, long before the existence of the decimal system, computations were often made in fractions which were multiples of $\frac{1}{100}$. These calculations in multiples of $\frac{1}{100}$ were used to figure out taxes. This is the same as calculating percentages. After a few centuries, this became the norm and came into common practice, so much so that this began to be taught in classrooms as well.

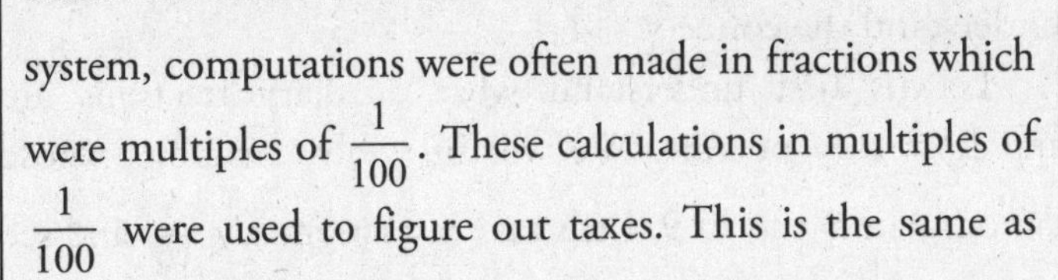

In this chapter, I will give a brand new treatment to percentages. The objective of this chapter is to offer you skills by which you will be able to convert any fraction into a percentage. So, we will learn how to convert fractions into percentages using a new method. And thereafter we will see, not only how to convert simple fractions into percentages, like $\frac{1}{2} = 50\%$ or $\frac{1}{5} = 20\%$, but we will also learn how to convert difficult fractions to percentages like $\frac{1}{7} = 14.28\%$ or $\frac{1}{13} = 7.69\%$.

The secret to converting fractions into percentages lies in a little known concept called 'a*uxiliary fractions' (AF).* Auxiliary means providing supplementary or additional help and support. So, with the support of auxiliary fractions, we should be able to achieve our objective of converting fractions into decimals and then, finally percentages. Remember, you will not find this in any school or college textbook—it will be a new thing, so be patient and try to understand the concept.

Let us first understand what auxiliary fractions are and how they operate. We will first solve fractions whose denominators end in 9. Let's take a fraction, say $\frac{5}{19}$ and see how to convert it into percentage.

For the fraction $\frac{5}{19}$, note that 19 is closest to 20. So, the auxiliary fraction is arrived at by dividing the numerator 5 by 20. This gives us the auxiliary fraction of $\frac{0.5}{2}$. We will now convert $\frac{0.5}{2}$ into our desired result like this.

Step 1

We divide 0.5 by 2, which gives us the quotient 0.2 and remainder 1. We prefix the remainder 1 just before 2, making it 12. Our example looks like this now:

$$\frac{5}{19} = 0._12$$

Step 2

Now, that we have 12, we divide it by 2, giving us our next digit 6 and 0 as remainder. We prefix the remainder 0, just before 6—making it 06. Our sum looks like this now:

$$\frac{5}{19} = 0._12_06$$

Step 3

So, now we have 06. We divide 06 by 2 which gives us 3 and the remainder 0, which we again prefix to 3. Now, our example looks like this:

$$\frac{5}{19} = 0._12_06_03\ldots$$

Step 4

Now that we have up to three decimals of $\frac{5}{19}$ we can multiply it by 100 to convert into a percentage. Our answer becomes $\frac{5}{19} = 26.3\%$ ·

You can check your answer with a calculator and be doubly sure. This method works with those denominators which end in a 9. Let's do couple of more examples and thereafter we shall see how to tackle other numbers where the denominators do not end in 9.

Let us now convert $\frac{8}{29}$ into percentage. Note that the denominator here ends in 9. So, we will convert $\frac{8}{29}$ into an auxiliary fraction. The number 29 is closest to 30, so that gives us $\frac{8}{29} \Rightarrow \frac{8}{30} \Rightarrow \frac{0.8}{3} AF$

Let's start our process.

Step 1

We divide 0.8 by 3 which gives us quotient of 0.2 and remainder 2, which we prefix to 2. Our sum looks like this:

$$\frac{8}{29} \Rightarrow \frac{0.8}{3} AF = 0._2 2$$

Step 2

We now divide 22 by 3, giving us 7 and remainder 1. We prefix 1 before 7. Our sum looks like this:

$$\frac{8}{29} \Rightarrow \frac{0.8}{3} AF = 0._2 2_1 7$$

Step 3

We continue the same step. So, we divide 17 by 3, which gives us 5 and remainder 2 which we prefix just before 5. It looks like this:

$$\frac{8}{29} \Rightarrow \frac{0.8}{3} AF = 0._{2}2_{1}7_{2}5$$

Let's go one more decimal point this time.

Step 4

We divide 25 by 3 giving us 8 and remainder 1. Like before, we prefix 1 to just before 8. The sum looks like this now:

$$\frac{8}{29} \Rightarrow \frac{0.8}{3} AF = 0._{2}2_{1}7_{2}5_{1}8\ldots$$

Step 5

Now we multiply our answer 0.2758 with 100 to convert it into a percentage. Our final answer is:

$$\frac{8}{29} = 27.58\%$$

Let's take another example to better understand this new method. Say we have $\frac{7}{39}$. Note here that the denominator ends in 9. So, we will convert $\frac{7}{39}$ into an auxiliary fraction. The number 39 is closest to 40, so our auxiliary fraction would be $\frac{7}{39} \Rightarrow \frac{7}{40} \Rightarrow \frac{0.7}{4} AF$. Now let us start our process.

Step 1

We divide 0.7 by 4. We get 0.1 as our quotient and 3 as remainder. We prefix the remainder 3 to just before 1. Our sum looks like this now.

$$\frac{7}{39} \Rightarrow \frac{7}{40} \Rightarrow \frac{0.7}{4} AF = 0._31$$

Step 2

We now divide 31 by 4. We get quotient of 7 and remainder 3 again which we prefix to 7. Our sum looks like this:

$$\frac{7}{39} \Rightarrow \frac{7}{40} \Rightarrow \frac{0.7}{4} AF = 0._31_37$$

Step 3

We now divide 37 by 4. It gives us 9 as quotient and remainder 1. We prefix 1 to 9 like this:

$$\frac{7}{39} \Rightarrow \frac{7}{40} \Rightarrow \frac{0.7}{4} AF = 0._31_37_19 \ldots$$

Step 4

Finally, now we multiply our answer 0.179 with 100 to convert it into a percentage. Our answer is 17.9%

$$\frac{7}{39} = 17.9\%$$

Converting Fractions into Percentages

Let's us now come to the point where we convert fractions starting from $\frac{1}{2}$ onwards into percentages. Let us look at the logic and the mathematical reasoning behind this.

So, we have:

$$\frac{1}{2} = 50\%$$

$$\frac{1}{3} = 33.3\%$$

So, the question now is that how do we calculate $\frac{2}{3}$? Well, we just double $\frac{1}{3}$ and hence it will be the double of 33.3%, which gives us 66.6% as our answer. $\frac{2}{3} = 66.6\%$. Hope this is clear.

Now we come to $\frac{1}{4} = 25\%$

Now $\frac{2}{4}$ is nothing but $\frac{1}{2}$, so $\frac{1}{2} = 50\%$

Now $\frac{3}{4} = \frac{1}{4} + \frac{2}{4} = 25\% + 50\% = 75\%$. Easy?

Let's now come to $\frac{1}{5}$.

$\frac{1}{5} = 20\%$ It's very simple and straightforward.

Now $\frac{2}{5} = 40\%$. We just double $\frac{1}{5}$ to arrive at the percentage of $\frac{2}{5}$. Hope this logic is clear.

Similarly, for $\frac{3}{5}$, we just need to add $\frac{1}{5}$ and $\frac{2}{5}$. So, we have 20% + 40%, giving us 60%. There can be multiple ways to arrive at our answer.

Let us now see $\frac{4}{5}$. This would be nothing but double of $\frac{2}{5}$. So $\frac{2}{5} = 40\%$.

So $\frac{4}{5}$ will be 40% + 40% = 80%.

Our answer is $\frac{4}{5} = 80\%$.

Let's come to $\frac{1}{6}$ now. Note that $\frac{1}{6} = \frac{1}{2} \times \frac{1}{3}$. So, since $\frac{1}{2}$ is 50%, we just divide 50% by 3. This gives us 16.6%. Our sum looks like this now $\frac{1}{6} = \frac{1}{2} \times \frac{1}{3} = \frac{50\%}{3} = 16.6\%$. Hope the logic is clear.

Now we come to $\frac{2}{6}$. That's nothing but $\frac{1}{3}$. So, we have $\frac{2}{6} = \frac{1}{3} = 33.3\%$

Now $\frac{3}{6}$ is the easiest. $\frac{3}{6}$ is equal to half. So, it is 50%.

Our fraction looks like $\frac{3}{6}=\frac{1}{2}=50\%$ We now come to $\frac{4}{6}$, which when simplified, gives us $\frac{2}{3}$ which is nothing but 66.6%.

And finally, we come to $\frac{5}{6}$. So, one way to calculate this would be $1-\frac{1}{6}=\frac{5}{6}$. We know 1 is nothing but 100% and from it we take away $\frac{1}{6}$ (16.6%). So, it looks like $100\%-16.6\%=83.3\%$, our answer.

Let us now see $\frac{1}{7}$. We will use the concept of auxiliary fractions now. Can you tell me how we can convert our denominator 7 to end in a nine? (Remember, we need the denominator to end in a 9 for us to be able to use the auxiliary fractions). So, to make the denominator end in 9, we multiply the denominator by 7. Why 7? Because 7 multiplied by 7 gives me 49, so we can then apply the rule. Also, now that we have multiplied the denominator by 7, we must multiply the numerator by 7 as well. So, we get $\frac{1}{7}=\frac{1\times 7}{7\times 7}=\frac{7}{49}$. Now we need our auxiliary fraction, which is nothing but $\frac{7}{49}\Rightarrow\frac{7}{50}\Rightarrow\frac{0.7}{5}AF$. Now let's start computing it.

Step 1

We divide 0.7 by 5. It gives me quotient digit of 1 and remainder 2. We prefix the remainder 2 just before to the quotient 1. Our sum looks like this:

$$\frac{1}{7} \Rightarrow \frac{0.7}{5} AF \Rightarrow 0._{2}1$$

Step 2

We now divide 21 by 5. It gives me the next quotient digit of 4 and remainder 1. Again, we prefix our remainder 1 to just before 4. Our sum looks like:

$$\frac{1}{7} \Rightarrow \frac{0.7}{5} AF \Rightarrow 0._{2}1_{1}4$$

Step 3

We now divide 14 by 5. We get 2 as our quotient digit and remainder as 4. We prefix 4 our remainder to just before 2.

$$\frac{1}{7} \Rightarrow \frac{0.7}{5} AF \Rightarrow 0._{2}1_{1}4_{4}2_{2}8_{3}5$$

Step 4

Now our answer looks like:

$$\frac{1}{7} \Rightarrow 0.1428 \times 100 = 14.285\%$$

We now come to $\frac{2}{7}$. It is simply the double of $\frac{1}{7}$. So we have

$14.285\% \times 2 = 28.57\%$. That's our answer.

So now its turn of $\frac{3}{7}$. Break $\frac{3}{7}$ like

$\frac{3}{7} = \frac{2}{7} + \frac{1}{7} = 28.57\% + 14.285\% = 42.85\%$

So similarly $\frac{4}{7} = 2 \times 28.4\% = 56.8\%$

Now can you calculate $\frac{5}{7}$ and $\frac{6}{7}$ in similar ways?

We now begin the computation of $\frac{1}{8}$.

Now $\frac{1}{8}$ can be expressed as the product of $\frac{1}{8} = \frac{1}{2} \times \frac{1}{4}$.

So, we just have to divide 50% by 4.

So that is $\frac{1}{8} = \frac{1}{2} \times \frac{1}{4} = \frac{50\%}{4} = 12.5\%$.

So $\frac{1}{8}$ Is 12.5%. Let's look at the other fractions.

For $\frac{2}{8}$, we can simplify further to get $\frac{2}{8} = \frac{1}{4} = 25\%$.

Our answer is 25%.

For $\frac{3}{8} = \frac{1}{8} + \frac{2}{8} = 12.5\% + 25\% = 37.5\%$.

$\frac{4}{8}$ Is the simplest. $\frac{4}{8} = \frac{1}{2} = 50\%$.

Building on from here, we arrive at $\frac{5}{8}$.

It is nothing but $\frac{5}{8} = \frac{4}{8} + \frac{1}{8} = 50\% + 12.5\% = 62.5\%$.

$\frac{6}{8}$ is simplified to $\frac{3}{4}$ which is 75%; We now come to $\frac{7}{8}$.

We can compute this in many ways. One of which will be to add $\frac{1}{8}$ to $\frac{6}{8}$ giving us 12.5% + 75% = 87.5%.

Let's now take $\frac{1}{9}$. We can use the concept of auxiliary fractions to solve it, since the denominator ends in a 9. So, we have $\frac{1}{9} \Rightarrow \frac{1}{10} \Rightarrow \frac{0.1}{1} AF$. Now let's start computing it.

Step 1

We divide 0.1 by 1. It gives us quotient of 1 and remainder 0. We prefix this remainder 0 to just before 1. It looks like this:

$= 0._{0}1$

Step 2

We again divide 01 with 1. It gives us 1 as quotient and 0 remainder which is prefixed to just before 1.

$= 0._{0}1_{0}1$

Step 3

It is a recurring answer. The quotient repeats. So our answer becomes:

$$\frac{1}{9} \Rightarrow \frac{1}{10} \Rightarrow \frac{0.1}{1} AF \Rightarrow 0._{0}1_{0}1_{0}1_{0}1$$

$$\frac{1}{9} = 0.11111 = 0.11111 \times 100 = 11.11\%\cdot$$

Now, getting the rest of the fractions is simple.

$$\frac{2}{9} = \frac{1}{9} \times \frac{2}{1} = 11.11\% \times 2 = 22.22\%$$

$$\frac{3}{9}=\frac{1}{9}\times\frac{3}{1}=11.11\%\times 3=33.33\%$$

So now you can compute from $\frac{4}{9}$ to $\frac{8}{9}$. Easy, isn't it?

Some Clues

Let me give you clues for computing some fractions where denominator is a prime number to a percentage, using auxiliary fractions. For example:

$\frac{1}{11}$ We can compute it into an auxiliary fraction.

We'll have $\frac{1}{11}\Rightarrow\frac{9}{99}\Rightarrow\frac{9}{100}\Rightarrow\frac{0.9}{10}AF$

Similarly, $\frac{1}{13}$ can also be converted into an auxiliary fraction.

So, it looks like $\frac{1}{13}\Rightarrow\frac{1\times 3}{13\times 3}\Rightarrow\frac{3}{39}\Rightarrow\frac{3}{40}\Rightarrow\frac{0.3}{4}AF$

And finally, $\frac{1}{17}$. We'll have:

$$\frac{1}{17}\Rightarrow\frac{1\times 7}{17\times 7}\Rightarrow\frac{7}{119}\Rightarrow\frac{7}{120}\Rightarrow\frac{0.7}{12}AF$$

And once we have the auxiliary fractions, converting it to percentage form is easy enough!

9

'Square Root' Adventures

In this chapter, we are going to take a look at how to calculate square roots of perfect squares.

First let's look at the list of numbers from 1 to 9 and their squares, last digits, and digit sums.

Number	Square	Last Digit	Digit Sum
1	1	1	1
2	4	4	4
3	9	9	9
4	16	6	7
5	25	5	7
6	36	6	9
7	49	9	4
8	64	4	1
9	81	1	9

Looking at the above given table, we can note that:

- The square numbers only have a digit sum of 1, 4, 7 and 9, and
- They only end in 1, 4, 5, 6, 9 and 0.

So, based on these points, it is simple to find out if a given number is a perfect square or not!

Let's take an example. Say we have to find the square root of $\sqrt{3249}$

Step 1

We first make pairs of 3249. Since there are two pairs (32 and 49) we can easily say that the number of digits in our answer will also be two.

Step 2

Now we'll look at the first two digits, 32, and notice that 32 is more than 25 (5^2) and less than 36 or 6^2. So, the answer will be between 50 and 60. This is because 50^2 is 2500 and 60^2 is 3600. Therefore, the square root of 3249 will be between 50 and 60.

Step 3

In our third step, we'll focus on the last figure of 3249, which is 9. Any number ending with 3 will end with 9, when it is squared, so the number we are looking for could be 53 but here the number could also be 57.

So, what is our answer? Is it 53 or 57?

Think about it for a moment!

Step 4

Now, we can use the digit sums to find out our answer.

So, if $53^2 = 3249$, then converting it into digit sums we get $(5 + 3)^2 = 64 = 6 + 4 = 10 = 1 + 0 = 1$

Hence it can't be 53 because the digit sums do not match.

So, our answer has to be 57.

We'll now check it again
$5 + 7 = 12 = 1 + 2 = 3$.
$3 \times 3 = 9$, therefore the digit sum of 3249 must be 9. And when we check, we find $3 + 2 + 4 + 9 = 9$.

Therefore, our final answer becomes 57.

Let's take another example. Say we have to find the square root of $\sqrt{2401}$

Step 1
As we did earlier, we'll first make pairs of 2401. Since there are two pairs, we can now say that the number of digits in our answer will also be two.

Step 2
Now looking at the first two digits, 24, we can see that 24 is more than 16 (4^2) and less than 25 or 5^2, so the answer will be between 40 and 50.

Step 3
We'll now look at the last digit in 2401, which is 1.
So, the sum could be either 41 or 49, as both squared will give the end digit as 1. This is because the last digit of 41^2 will be 1×1 which equals 1 and the last digit of 49^2 will be $9 \times 9 = 81$. Hope this is clear.

Step 4
Now in our final step, we'll use digit sums to get our final digit.

So, the digit sum of 41^2 is 7 and the digit sum of 49^2 is also 7. Now what do we do with it?

We need to find the square of 45, which is very easy to find and that is, 2025.

We choose 45^2 because it is in the middle of 41^2 and 49^2.

Now since 49 is more than 45, our answer becomes 49^2.

This method of finding the square roots is extremely easy and in fact, can be done mentally!

Let's take another example and this time, we'll try to do it mentally.

Say we need to find the square root of $\sqrt{9604}$.

Step 1

We'll first make pairs and there are two pairs here, which means that the answer will have 2 digits.

Step 2

Now, we'll take the first pair and the perfect square just less than 96 is 81 (9^2) and the perfect square just more than 96 is 100 (10^2)

So, our answer must be between 90 and 100.

Step 3

In the final step, we'll take 9604, and here the last digit is 4, so the square root can be either 92 or 98.

We'll then check it using digit sums and we get our square root to be 98. That's all folks, it's that easy!

Now let's take a slightly bigger example. Say we have to find the square root of $\sqrt{24964}$

Step 1

So, first we make pairs in our given sum. The pairs are 02, 49 and 64. Please note that there are three pairs now, therefore there'll be 3 digits in our final answer.

Step 2

Note that since 249 lies between 15^2 and 16^2, the first two figures must be 15. We choose 15^2 because 15^2 is 225 and 16^2 is 256. Since we have to find the root of 24964, for the first two digits we will choose 15^2 because it is less than 249.

Step 3

And in our final step, we'll consider the last figure of our sum, that is 4. So, the answer can be either 152 or 158. These are the two possibilities.

So, we'll check by digit sum and the answer that we get is 158.

Now let's take a final example. Say, we have to find the Square Root of $\sqrt{32761}$

Step 1

So, we'll first make pairs in our given sum. Here we see that there are three pairs, therefore there will be 3 digits in our final answer.

Step 2

Now, since 327 lies between 18^2 and 19^2, the first two figures must be 18.

Step 3

As the last figure of our problem is 1, the answer can be either 181 or 189. These are the two possibilities, so it's time to check it using the digit sum and the answer that we get is 181. So, the square root of 32761 is 181.

10

Fastest Finger First!

This final chapter is for those who have difficulty remembering their times tables or number bonds. This is a worldwide problem and it's not your fault if you do not know the times tables yet! Maths shouldn't be about rote learning all the time. I will give you a very tactile way to do it—right on your fingers! So, let's get started with the 9 times tables first!

9x Times Tables on Your Fingers

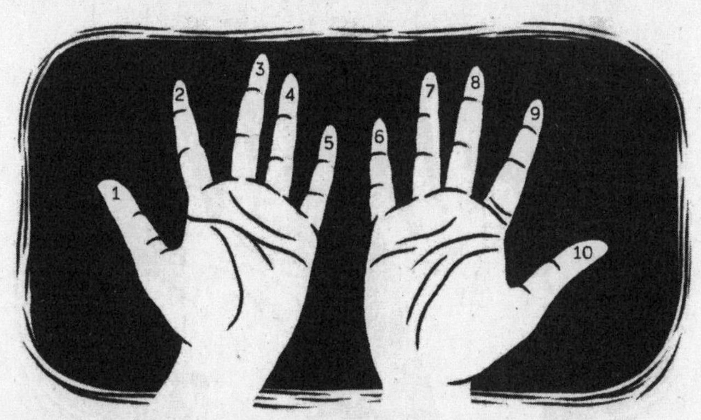

First count from the left thumb rightwards, numbering each finger 1 to 10 as you go.

Now to do the 9 times tables you need to fold the finger associated with the number, nine is to be multiplied with.

So, to do 9 × 1, you simply fold the first finger.

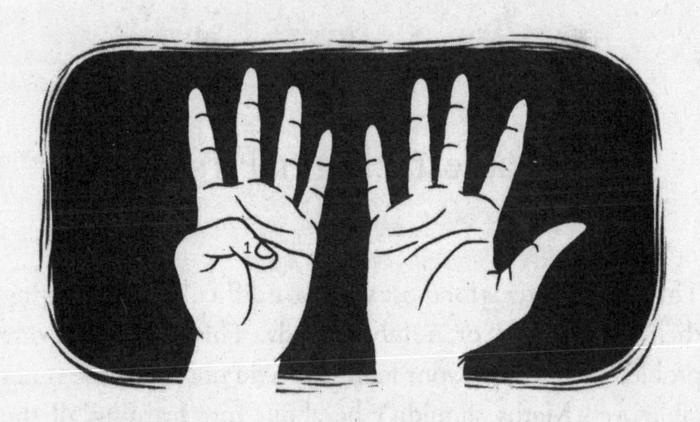

You then just count the remaining fingers. You see there are nine fingers remaining, so your answer for 9 × 1 is 9.

Now you have to do 9 × 2. For that, fold your second finger.

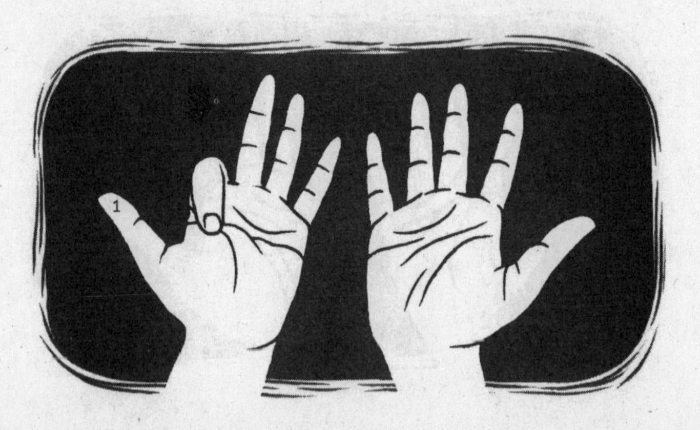

Now you see, to the left of the folded finger there is one finger and to the right of the folded finger, you can count eight fingers. You combine 1 and 8 together to give you 18. This is your answer to 9 × 2. Yes, it's that easy!

Let's now see 9 × 3.

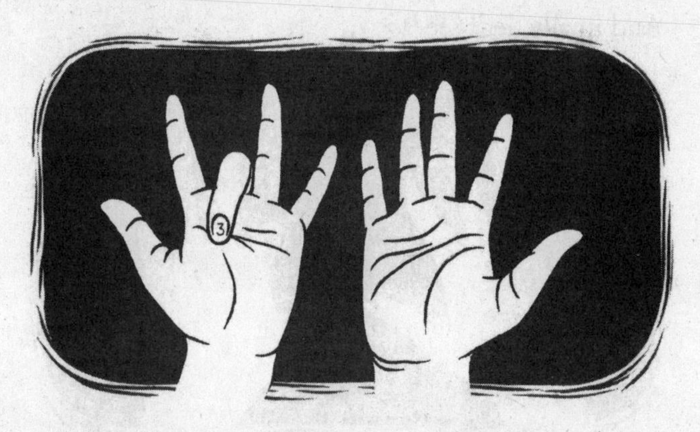

Similarly, we fold the third finger now. We see that, to the left on the folded finger we have 2 fingers and to the right of the folded finger, we can count 7 fingers. We combine 2 and 7 to give 27 as our answer!

Let's try 9 × 5.

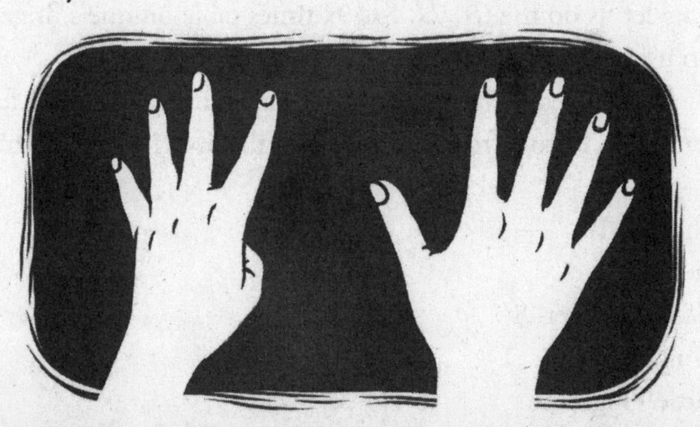

We reverse our hands as it may be slightly uncomfortable folding some fingers. The same logic applies. Here we fold the fifth finger. We see that we have 4 fingers to the left of the folded finger and 5 fingers to the right of the folded finger. We combine 4 and 5 to get 45 as our answer.

And finally, let's see 9 × 7.

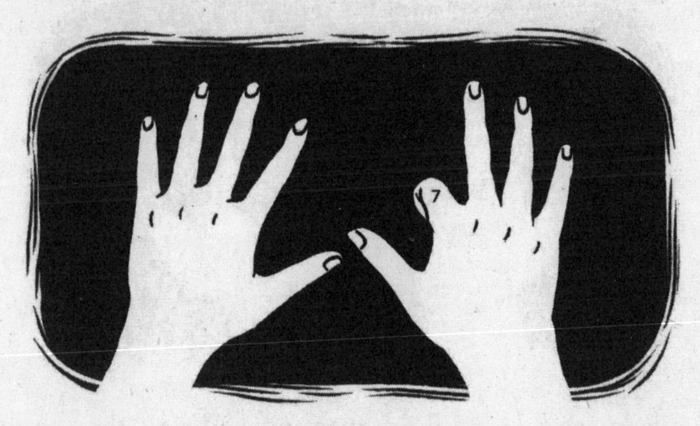

We have 6 fingers to the left of our folded seventh finger and 3 fingers to the right. Combining, we have 63 as our answer.

The Universal Times Tables

Now let us do the 6x, 7x, 8x, 9x times table on one's fingers too using another new method.

Number your fingers on both hands using coloured felt pens from 10 to 6 from the thumb to the little finger like this:

Thumb–10
Index–9
Middle Finger–8
Ring Finger–7
Little Finger–6

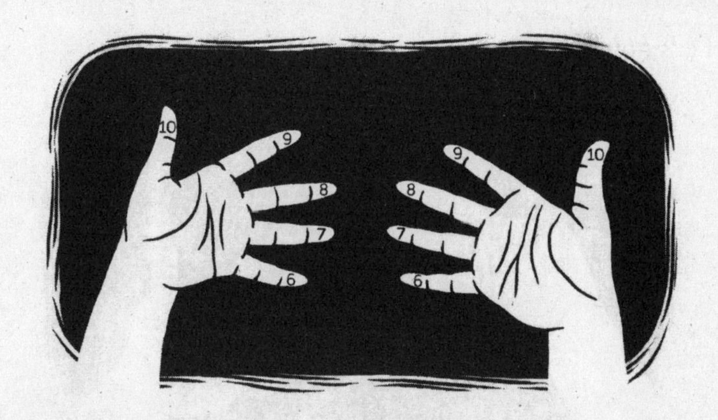

Now say you would like to do 9 times 8.

So, you join the 9 on your left hand to the 8 on your right hand, as shown in the picture below.

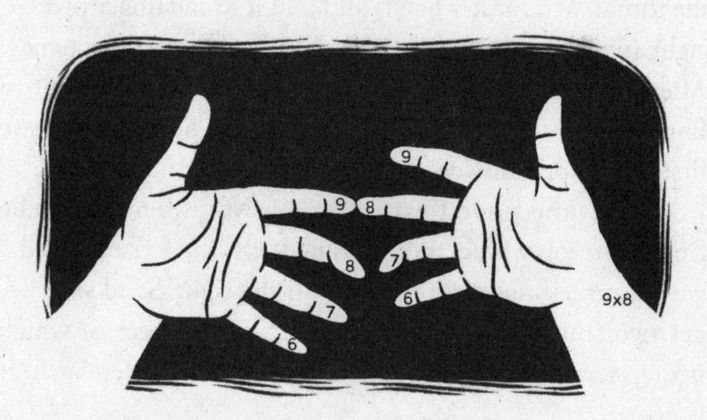

After joining the fingers as above, you count the joined fingers and the fingers below the jointed portion. So, you have seven fingers below including the joined fingers. On top, you have one finger on the left-hand side and two fingers on the right-hand side. So, you multiply 1 with 2 to get 2. Therefore, we have 7 and 2, 72 as our answer.

Let's now try 7 times 8

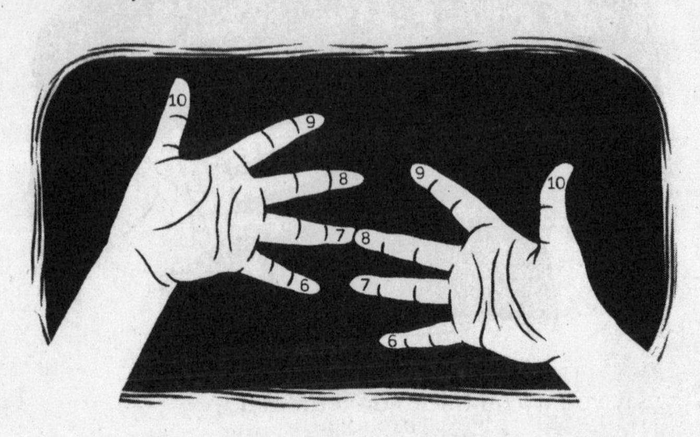

So, you take the finger marked seven (which is fourth from the thumb of your left hand) and join it to the finger marked eight (which is third from the thumb of your right hand). After joining the two fingers, you count them. You get five fingers below (including the joined fingers). So that forms the first part of our answer—5.

To get the second part of the answer, we multiply the fingers on top which are not joined. On the left-hand side, we get three fingers on top and on the right-hand side, we get two fingers. So, we multiply 2 and 3 to get 6, which is the second part of our answer. So, our complete answer is 56.

Let's try a carry-over sum now. Say, we want to do 6 times 6 now. So, we see both our little fingers joined this time.

The two joined little fingers give us 2 and it represents two tens which is nothing but 20.

And we see that on top, we have 4 fingers on each side. So, we multiply 4 with 4 to get 16. Now, to get our answer, we add 20 to 16. This gives us 36.

Let's see some more examples. How about 8 × 8?

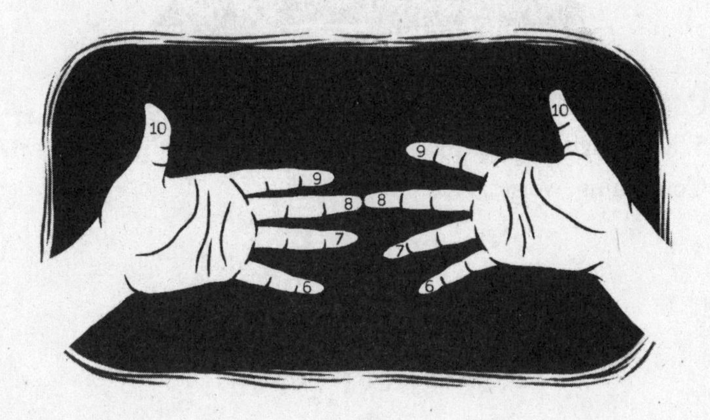

The fingers marked 8 are joined. The joined fingers and the fingers lying below them are counted. We get 6. The unjoined fingers on top are counted on both the hands. On the left hand, we get 2 and on the right hand we get 2 as well. We multiply 2 with 2 to get 4. We combine 6 and 4 to get 64 as our answer.

Let's finally do 6 × 9 on our fingers.

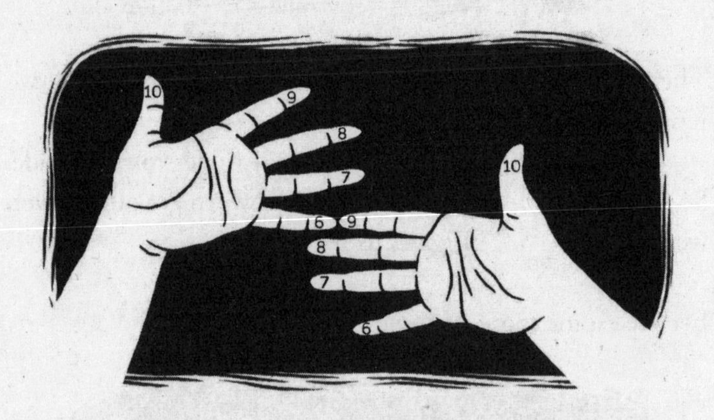

Counting the joined fingers and the fingers below them, we get 5. Multiplying the fingers above in both hands we get 4 × 1 = 4. Combining, we get 54 as our answer!

Notes and Credits

Chapter 1: All About 'Additions'

Mental additions are important as they form a robust foundation for all our other calculations. I felt thorough joy in sharing some maths sutras on mental addition in this book.

- Maharaja, Jagadguru Swami Sri Bharati Krishna Tirthaji. *Vedic Mathematics: Sixteen Simple Mathematical Formulae from the Vedas*. Motilal Banarsidass Publishers, 1965.
- Tekriwal, Gaurav. *Maths Sutra: The Art of Vedic Speed Calculation*. Penguin Random House India, 2015.

I would also like to credit the following book which forms the base of the short micro-story on Jakow Trachtenberg:

- Trachtenberg, Jackow. *The Trachtenberg Speed System of Basic Mathematics*, trans. Ann Cutler. Souvenir Press Ltd., 1960.

For more practice on Kakuro the Japanese puzzle, visit: www.kakuroconquest.com and www.kakuros.com.

Chapter 2: The 'Difference' Matters

B.Sai Kiran attempted a Guinness record in subtracting a 70-digits number from another 70-digit number in 60.05 seconds. He used the method 'super subtraction' as shown in this book. You will find more on him in this article:

- *The Hindu*. 'Numerical prodigy sets Guinness record.' 10 April 2012, Hyderabad edition. http://www.thehindu.com/news/cities/Hyderabad/numerical-prodigy-setsguinness-record/article3299491.ece.

I am thankful to Robert Fuhrer of KenKen Puzzle, LLC, New York for giving me permission to write about and use KenKen puzzles in this book for my readers. KenKen® is a registered trademark of Nextoy, LLC.

Watch the Talks at Google session on *KenKen: Happiness Through Math* given by Tetsuya Miyamoto, founder of KenKen in 2016. Try to understand his philosophy, 'The art of teaching without teaching'. You can find the video here:

- 'Tetsuya Miyamoto "KenKen: Happiness Through Math" | Talks at Google.' Talk by Tetsuya Miyamoto. YouTube video. Posted by 'Talks at Google', 28 July 2016. https://www.youtube.com/watch?v=gh4YV3VvmEE.

Chapter 3: Speed up Your 'Multiplication'

Saint Tirthaji's work cited under no.1 above is the passion and inspiration behind this chapter on multiplication.

I would like to recommend my readers to also go through the works of Arthur T. Benjamin, American mathematician and author, who is also a noted TED speaker.

- Benjamin, Arthur T. and Michael Shermer. *Secrets of Mental Math: The Mathemagician's Guide to Lightning Calculation and Amazing Math Tricks.* RHUS Publishers, 2006.
- Benjamin, Arthur T. *The Magic of Math.* Basic Books, 2015.

Chapter 4: 'Divide' like an Egyptian

Michael S. Schneider's YouTube video on Egyptian Maths is the inspiration behind this chapter to simplify division! You can look it up here:

- 'Egyptian Maths.' Talk by Michael S. Schneider. YouTube video. Posted by 'mothnrust', 5 September 2008. www.youtube.com/watch?v=Ih1ZWE3pe9o&t=7s.

I am thankful to Mr Daniel Brillon, director of operations at Singapore Math Inc for giving me permission to write about Singapore Math.

Singapore Math® is a registered trademark of Singapore Math Inc and Marshall Cavendish Education Private Limited.

An alphametic puzzle is also sometimes known as a cryptarithm. And if you get stuck at any certain alphametic, visit:

- Alphametic Puzzle Solver. Copyright 1998 by Truman Collins, 1998. Last accessed 29 November 2017. www.tkcs-collins.com/truman/alphamet/alpha_solve.shtml.

Chapter 5: The 'Digit Sum' to Check Your Answers

I am thankful for the Vedic square activity which was prepared by David Benjamin at www.teachitmaths.co.uk.

In October 2009, I came across a pattern of digit sums called One Eye. You can work out the Vedic square in a spreadsheet and see the beautiful pattern which emerges. For more, visit:

- Vedic Maths India. Last accessed 29 November 2017. www.vedicmathsindia.org/blog/2009/10/oneeye/.

Chapter 6: No More Fear of 'Fractions'

I am thankful to Prof William Jackson from www.mathdemystified.com for his permission to share some of his work on Singapore Math with my readers, as detailed under Chapter 4 above. You can also see the article 'Egyptian Fraction'. Find it here:

- Wolfram MathWorld. Last accessed 29 November 2017. http://mathworld.wolfram.com/EgyptianFraction.html.

Chapter 7: The Magic of Magic Squares!

Image Credit of the Magic square at the Parshvanath temple, Khajuraho: Dr Rainer Typke.

Chapter 8: Destination 'Percentages!'

In this chapter, I have given a very extensive treatment to percentages. This has not been seen before. I have used Tirthaji's concept of auxiliary fractions to convert fractions into decimals and finally decimals into percentages. For references to Tirthaji's and Gaurav Tekriwal's books, see Chapter 1 above.

Chapter 9: 'Square Root' Adventures

- Tekriwal, Gaurav. *Maths Sutra: The Art of Vedic Speed Calculation*. Penguin Random House India, 2015.

Chapter 10: Fastest Finger First!

There are other methods of finger counting available as well. Look up the Korean finger counting method called Chisanbop.

- Pai, S.J. and H.Y.Pai. *Complete Book of Chisanbop: Original Finger Calculation Method.* Van Nostrand Reinhold Inc., U.S., 1981.

Acknowledgements

First, I would like to thank all the readers of my previous book *Maths Sutra: The Art of Vedic Speed Calculation* for making the book a grand success and sending me emails of appreciation. Those small notes of encouragement led me to shape and structure this reference book—*Maths Sutras From Around the World*—for younger learners.

I would like to thank Mr Robert Fuhrer from KenKen Puzzle LLC, New York; Mr Daniel Brillon, director of operations at Singapore Math Inc. and Prof. William Jackson from mathdemystified.com for their support and permission to let more people become aware of their innovative methods and puzzles.

I would like to thank Ms Sohini Mitra at Penguin Random House India for believing in me and being patient with me for this book. She showed me her vision and I hope this book truly resonates with younger learners and young adults. I would also like to thank Mr Jit Chowdhury for bringing this book of numbers to life with his amazing illustrations. I am grateful to Mr Soumil Roy for the book website.

On the personal front, I would like to thank my parents—especially my mother Mrs Madhu Tekriwal for her encouragement and inspiration. I am grateful to my wife Shree for managing the office on my behalf and helping me with my midnight food cravings while I worked on the book. Miraaya, my three-year-old, needs to be thanked as well for her love and for simply being cute all the time.